M 95 GU

22-3-77

KU-309-356

B
1:
.5
.GUN

# Pesticides and human welfare

Social Science Library
Oxford University Library Services
Manor Road
Oxford OX1 3UQ
WITHDRAWN

WITHDRAWN

301047967.

This book has been sponsored by the following companies:

BASF AG, Federal Republic of Germany

Bayer AG, Federal Republic of Germany

Ciba-Geigy AG, Switzerland

Hoechst AG, Federal Republic of Germany

Imperial Chemical Industries Limited, England

Rhône-Poulenc SA, France

Shell International Chemical Company Limited, England

# Pesticides and human welfare

EDITED BY
D. L. GUNN
AND
J. G. R. STEVENS

OXFORD UNIVERSITY PRESS
1976

*Oxford University Press, Walton Street, Oxford* OX2 6DP

OXFORD  LONDON  GLASGOW  NEW YORK
TORONTO  MELBOURNE  WELLINGTON  CAPE TOWN
IBADAN  NAIROBI  DAR ES SALAAM  LUSAKA  ADDIS ABABA
KUALA LUMPUR  SINGAPORE  JAKARTA  HONG KONG  TOKYO
DELHI  BOMBAY  CALCUTTA  MADRAS  KARACHI

Casebound ISBN 0 19 854522 3
Paperback ISBN 0 19 854526 6

© Oxford University Press 1976

*All rights reserved. No part of this publication may be reproduced,
stored in a retrieval system, or transmitted, in any form or by any
means, electronic, mechanical, photocopying, recording or otherwise,
without the prior permission of Oxford University Press*

UNIVERSITY OF OXFORD

INSTITUTE OF AGRICULTURAL

ECONOMICS

DATE CATALOGUED 2|5|77

CLASS-MARK

WITHDRAWN

ON PERMANENT LOAN FROM CBAE.

Printed in Great Britain
by Thomson Litho Ltd., East Kilbride.

# Preface

During the past ten or fifteen years, public interest in the environment and in the welfare of wild life and domestic animals, as well as of man himself, has grown remarkably in the advanced countries. Much of this growth has been the result of more leisure among townspeople and greater prosperity generally in these countries. It has been intensified by propaganda, some of which opposes certain modern farming methods, including various forms of intensive livestock production and also the use of chemicals – particularly pesticides. Although much of the recent criticism of farming practices has been exaggerated and unsound it has, inevitably, aroused the sympathy of many well-intentioned people.

Against this, it is widely recognised that since the modern synthetic pesticides started to come into use about thirty years ago, they have made substantial contributions to the agriculture of both the developed and the developing countries. In the latter, especially, they have saved millions of human lives and alleviated distress in a great many people, as well as in domestic animals. In some of the less developed countries hunger and malnutrition are common among the very people, the small cultivators, whose business is to produce food; yet the introduction of modern techniques, the use of pesticides among them, could enable food to be produced and stored to feed them adequately.

There is no doubt that a good deal of confusion still exists among unbiased people about whether pesticides are 'a good thing', whose use should be encouraged and extended, or 'a bad thing', whose use should be curtailed or even stopped. In fact, of course, the matter is by no means a simple one of right or wrong, and it is in order to try to put the record straight, and to help people decide where the balance of advantage lies, that this book has been written. It has to be realised that a decision often involves weighing an advantage to human beings against an advantage to wild life.

In seeking authors to cover a very wide field, to which, we believe, no single writer could have done justice, we invited contributions from acknowledged experts. We chose those who could be relied upon to look at problems as a whole and to treat them judicially and accurately, in a way which would both satisfy a discriminating lay readership and also stand up to the scrutiny of scientific readers, or those with specialist knowledge.

Most of the authors are or have been on the staff of universities and government or international research institutes and agencies; and in two cases we have turned to authors in industry for subjects on which they were especially well qualified to write. We hope that the effect has been to produce a balanced and objective statement: this was certainly the brief we received from the seven companies, listed on page ii, who have sponsored this book, and who have allowed us complete editorial freedom.

The book is presented in three parts and an Appendix. The four chapters of Part I are to a greater or lesser extent introductory. They are written against the background of inadequate food supplies in some parts of the world, rapidly expanding populations, especially in those areas where food is most scarce, the need to control the incapacitating tropical diseases which afflict so many people in these same areas, and the financial problems of improving agriculture in both the developed and the developing countries.

Part II discusses what is being done to increase productivity by using pesticides in both temperate and tropical agriculture. The need to protect economically important crops against insect pests, diseases, and weeds is examined by specialists with knowledge of those crops, as is the importance of protecting livestock and stored products. In Part III the authors consider what some might call the 'black' or 'grey' areas of pest control (including the real and imagined hazards to people and to the environment, and the precautions that are taken to reduce dangers to a minimum), and evaluate (we hope realistically) non-chemical methods of controlling pests.

The Appendix provides not only 'a guide to terminology', but an introduction to pesticides which will prove invaluable to the layman: indeed, readers who have little knowledge of the subject may well find it useful to read the Appendix before reading the rest of the book. As we have already said, this book has been written for a wide readership; scientific and technical readers who have a particular interest in the subject will have no difficulty in following the authors' arguments and trains of thought. All readers will find that the opening and closing paragraphs of each chapter will tell them what the author is discussing, and its relevance to the book as a whole.

We are grateful to the authors, many of them distinguished, and almost all of them exceedingly busy men, for finding the time to contribute to this book — and often enough for their forbearance with the editors' whims. We owe a special debt of gratitude to the late Dr. J. M. Barnes, who completed his chapter just before his untimely death. We should like to thank the sponsors for asking us to undertake what has proved to be a particularly interesting and stimulating job, and the staff of the Oxford University Press for much invaluable advice on preparing the book for press.

D.L.G.
J.G.R.S.

# Abbreviations

Throughout the text the initials FAO have been used for the Food and Agricultural Organization of the United Nations, Rome, and WHO for the World Health Organization, Geneva. Abbreviations for titles of journals are in accordance with the fourth edition of the *World list of scientific periodicals* (Butterworth and Co. 1963-5).

# Contributors

Dr. A. V. ADAM
Pesticides Officer, FAO, Rome.

D. J. ANSELL, B.Sc.
Lecturer in Agricultural Economics, University of Reading, Berkshire.

†Dr. J. M. BARNES, C.B.E., M.B.
Late Director, M.R.C. Toxicology Unit, Medical Research Council
Laboratories, Carshalton, Surrey.

Dr. J. R. BUSVINE, F.I.Biol.
Professor of Entomology as applied to Hygiene, London School of
Hygiene and Tropical Medicine.

Dr. E. E. CHEESMAN, C.B.E.
Sometime Professor of Botany. Imperial College of Tropical
Agriculture, Trinidad. Lately a scientific adviser to the Secretary
of the Agricultural Research Council, London.

Dr. G. DAVIDSON
Reader in Entomology as applied to Malaria, Ross Institute of Tropical
Hygiene, London School of Hygiene and Tropical Medicine.

S. K. DE DATTA
International Rice Research Institute, Manila, Philippines.

I. D. FARQUHARSON, B.A.
Shell International Chemical Company Limited.

Dr. J. A. FREEMAN, O.B.E.
Pest Infestation Control Laboratory, Slough, Buckinghamshire.

Dr. W. R. FURTICK
Formerly Chief, Crop Protection Service, FAO, Rome.

K. S. GEORGE, B.Sc.,A.R.C.S.
Plant Pathology Laboratory, Hatching Green, Harpenden, Hertfordshire.

Dr. R. F. GLASSER
Overseas Services, Shell Oil Company, Houston, Texas.

Dr. R. H. GRAY
Department of Medical Statistics and Epidemiology, London School
of Hygiene and Tropical Medicine.

Dr. D. L. GUNN, C.B.E., F.I.Biol.
Lately a scientific adviser to the Secretary of the Agricultural
Research Council, London.

Dr. J. P. HUDSON, C.B.E., G.M., F.I.Biol.
   Editor, *Experimental Agriculture*. Lately Professor of Horticultural
   Science, University of Bristol, and Director, Long Ashton Research
   Station.

F. K. IMRIE, B.Sc.
   Philip Lyle Memorial Research Laboratory, Whiteknights, Reading.

Dr. D. E. JACOBS, M.R.C.V.S.
   Lecturer in Veterinary Parasitology, Department of Pathology, Royal
   Veterinary College, University of London.

Dr. D. PRICE JONES, F.I.Biol.
   Consultant biologist, Reading, Berkshire.

Dr. J. E. KING
   Plant Pathology Laboratory, Hatching Green, Harpenden,
   Hertfordshire.

Professor Dr. J. KRANZ
   Giessen University, Federal Republic of Germany.

G. A. MATTHEWS, B.Sc.
   Lecturer in Zoology, Imperial College of Science and Technology,
   Imperial College Field Station, Silwood Park, Ascot, Berkshire.

W. W. MAYNE, O.B.E., B.Sc., F.I.Biol.
   Sometime Coffee Scientific Officer, United Planters Association of
   Southern India. Lately General Manager of South Indian tea estates,
   James Finlay Group.

Dr. K. MELLANBY, C.B.E.
   Research Fellow, Natural Environment Research Council. Lately
   Director, Monks Wood Experimental Station, Nature Conservancy.

S. H. OU
   International Rice Research Institute, Manila, Philippines.

Dr. M. D. PATHAK
   International Rice Research Institute, Manila, Philippines.

Professor Dr. G. SCHUHMANN
   President, Federal Biological Institute, Braunschweig, Masseveg,
   German Federal Republic.

J. M. WALLER, M.A., D.I.C., Dip. Agric. Sci. O.D.M.
   Plant Pathology Liaison Officer, Commonwealth Mycological
   Institute, Kew, Surrey.

Dr. C. C. WEBSTER, C.M.G., F.I.Biol.
   Lately Chief Scientific Officer to the Agricultural Research Council,
   London.

# Contents

*Contents*

# PART I

# THE PROBLEMS

Plate 1.1. Fruit and vegetables in Bangkok's floating market. Intensification of food production in the tropics is possible where adequate water, fertility, and pest control can be provided. (FAO photograph by G. De Sabatino)

# 1 Uncontrolled pests or adequate food?

by William R. Furtick

Civilisation as it is known today could not have evolved, nor can it survive, without an adequate food supply. Yet food is something that is taken for granted by most world leaders despite the fact that more than half of the population of the world is hungry. Man seems to insist on ignoring the lessons available from history. The invention of agriculture, however, did not permanently emancipate man from the fear of food shortages, hunger, and famine. Even in prehistoric times, population growth often must have threatened or exceeded man's ability to produce enough food. Then, when droughts or outbreaks of diseases and insect pests ravaged crops, famine resulted.

That such catastrophies occurred periodically in ancient times is amply clear from the numerous biblical references. Thus, the Lord said: 'I have smitten you with blasting and mildew' (Amos 4:9); 'The seed is rotten under the clods, the garners are laid desolate, the barns are broken down; for the corn is withered . . . The beasts of the fields cry also unto thee: for the rivers of waters are dried up and the fire hath devoured the pastures of the wilderness' (Joel 1:17,20).

Plant diseases, drought, desolation, despair were recurrent catastrophies during the ages.

Thus spoke Norman Borlaug in opening his lecture on the occasion of being awarded the Nobel Peace Prize for 1970. His words turned out to be of short-range significance, for poor harvests in the Soviet Union and other areas of the world in 1972 and 1975 quickly caused most world leaders and the public at large to stop taking the food supply for granted. The high food prices which resulted had an impact on the families of both the more affluent and poorer sectors of the world population. All the world's housewives are now suffering from the little-known fact that each time they prepare a meal they have fed extra invisible guests who ate before them. These are the myriad crop pests that consumed from a quarter to one-half of the food before it ever reached the table. Neither the hungry nor the affluent can continue to pay this price, which is to receive only part of their daily bread.

The losses caused to our food supply by insects, crop diseases, rodents, weeds, and similar pests have recently gained the attention of world leaders in the highest councils. Dr. Henry Kissinger, the American Secretary of State, emphasized the magnitude of these losses and proposed urgent action to reduce them both in his statement which opened the United Nations World Food Conference in Rome in November 1974, and before a special session of the United Nations General Assembly in September 1975. This concern was echoed by large numbers of world leaders at the World Food Conference and led to a demand by many that an adequate supply of pesticides to combat these losses be assured. This was formalized by a special resolution on pesticides adopted by the World Food Conference. A special session of the United Nations General Assembly ten months later adopted a special resolution on reducing food losses after harvest. This resolution states:

> Developing countries should accord high priority to agricultural and fisheries development, increase investment accordingly and adopt policies which give adequate incentives to agricultural producers. It is a responsibility of each State concerned, in accordance with its sovereign judgement and development plans and policies, to promote interaction between expansion of food production and socio-economic reforms, with a view to achieving an integrated rural development. The further reduction of post-harvest food losses in developing countries should be undertaken as a matter of priority, with a view to reaching at least a 50 per cent reduction by 1985. All countries and competent international organizations should operate financially and technically in the effort to achieve this objective. Particular attention should be given to improvement in the systems of distribution of foodstuffs.

The rapid change in the world food supply between the over-abundance in the decades of the 1950s and 1960s to that of precariously low reserve stocks which followed the poor harvests of 1972 and 1975 resulted in a concerted international effort to increase world food production as an urgent priority. With new emphasis on production, there was a growing realization that protection not only of the present production but also of the future increased production from destruction by pests is a special imperative.

## The role of plant protection in increasing food production

Although the magnitude of losses from pests has not been adequately measured even in the most highly developed countries, these losses are recognized as being substantial. In developing countries, the pre-harvest and post-harvest crop losses are estimated by FAO to be in the region of 30 per cent or more of potential production. Even in the most highly developed agricultural countries they are still large. These are caused by diverse species of arthropod and vertebrate animals such as insects and rodents; weedy terrestrial and aquatic plants; plant diseases caused by bacteria, fungi, viruses, and micoplasm; and plant nematodes. These losses do not include the impact on food production caused by the low efficiency of agricultural workers, who suffer from various vector-borne diseases and internal and external parasites. The same factors are important in livestock production and cause major losses in the production of meat; and in some areas vector-transmitted diseases make large geographic areas unfit for certain types of livestock production. To these direct losses must be added the increased food costs to consumers, particularly in the rapidly expanding urban centres, who must pay higher prices as a result of the wastage and loss of production caused by pests.

Leading plant protection experts called together by FAO after the World Food Conference recognized that pesticides will during the foreseeable future remain a primary measure for combating losses from pests. With this in mind, it should be noted that in 1974 all developing countries combined used only about 10 per cent of the world pesticide production, and most of this was used in public health vector control programmes against flies and mosquitoes, and on export cash crops. Both the importance of pesticides for pest control and the need for their greater use in a proper manner if developing countries are to meet their food production needs are readily apparent. All the same, several complex problems are involved.

The control of pests with both pesticides and non-pesticide control measures requires a more highly developed infrastructure and better manpower training than is generally true for most other agricultural inputs. Without adequate pest control a large measure of the benefit from other inputs may be lost. These components of agricultural modernization must therefore proceed in harmony. This can be readily observed in developing countries, where until recently nearly all pest-icides used in agriculture have been on cash crops. As the high-yielding varieties of food crops have been introduced and used more widely, there has been a rapid increase in pesticide use on this small but increasing component of food crop production.

## Recent world food production trends

It would be somewhat simplistic to suggest that in the 1970s the world food situation suddenly took a wrong turn under the impact of bad weather. An analysis of production trends during the past few decades indicates that more deep-seated problems were also accumulating, with the industrialized countries producing more food than they could consume or export and the developing countries facing food import bills that were growing larger every year. The industrial countries could not sell all the food they were producing, and the developing countries could not produce fast enough to supply their needs.

Contrary to popular belief, the difference did not lie in the dynamics of agricultural production. The developing countries, in spite of their difficulties, were expanding their agricultural output in the 1950s and 1960s just as fast as the industrial countries—a truly remarkable achievement. The difference lay in the rates of food growth demand, which increased by 2.5 per cent per annum in the industrial countries as against 3.5 per cent in the developing countries, mainly because of faster population growth in the latter.

During the period following the Second World War, the performance of agriculture has differed greatly from country to country. In more than 20 of the developing countries the rate of growth of food production exceeded 4 per cent per annum, while in about a dozen countries it was less than 2 per cent during the period 1953-71. In about 30 developing countries food output grew faster than food demand, and in about 50 it grew faster than population. Even in the food crisis year of 1972 the developing countries' food production was 20 per cent higher than in 1966, the latest previous year of widespread bad weather, so that progress even between the troughs in the long-term trend had kept ahead of population growth.

These considerable achievements reflected dramatic and effective application of technology in many countries. However, it was not vigorous or effective enough, and this shows how enormous is the strain put upon the agricultural sector in countries whose population and general economic activity are both expanding rapidly. Not only was the effort insufficient to meet the rise in domestic food demand, with the consequence of large and ever larger food imports, but it also meant a disappointing performance in agricultural exports, which are the chief source of foreign exchange for most developing countries.

In addition to the accumulating problems of food production, there is the equally vital issue of the nutritional adequacy of supplies within countries and the extent of under-nutrition and malnutrition. Taking a conservative view, it would appear that out of nearly 100 developing countries, about 60 had a deficit on food energy supplies in 1970. In

6

the Far East and Africa, 25 per cent and 30 per cent of the population
is estimated to suffer from significant under-nutrition. Altogether in
the developing world (excluding the Asian centrally planned economies
for which insufficient information is available) malnutrition affects at
least 460 million people.

## The food problem of the future

Although the long-term average increase in world food production has
been greater than the growth of population ever since the Second World
War, the margin was smaller in the 1960s than in the 1950s. The
increase in food production slowed down in the 1960s in every major
region except Africa and North America. In part this reflects the large
element of post-war recovery in the early 1950s. In the industrial
countries population growth has declined, and part of the slower
increase in production in the 1960s was due to deliberate government
policies. In the developing countries, on the other hand, it occurred in
spite of accelerated population growth and government policies aimed
generally at increasing the rate of food production.

Thus, although total food production has increased at about the
same rate in the industrial and the developing countries, on a *per
caput* basis the increase has been much smaller in the latter. This has
meant that the already large difference in the actual level of production
per head between these two groups of countries has widened still
further: in the developing countries it was little more than one-quarter
of that in the industrial countries in 1971-3, as compared with about
one-third in 1961-3.

The fact that for so long a period food production in the developing
countries as a whole kept ahead of a rate of population growth
unprecedented in world history is a tremendous achievement. In many
individual countries, however, developments have been much less
favourable.

Domestic agricultural production is the main determinant of the
level of available food supplies in most developing countries, but it has
other important roles in their overall economic and social development,
including the earning and saving of foreign exchange, and the provision
of employment and of much of the capital needed for the development
of the rest of the economy. However, diversion of this capital from
agriculture may have contributed to reducing further the increase in
food production. The importance given to agricultural production is
reflected by the average targets in the various national development

plans being set at 4 per cent for the developing countries as a whole. Even so, only about one-third of the developing countries have reached their targets.

Increasing production depends basically on expanding the inputs of the different factors of production: land and water, labour, material inputs, the various types of capital, and technological know-how, including pesticide use. For farmers in developing countries, and especially the numerous small farmers, the possibility of using – and the incentive to use – more inputs in turn depends to a great extent on the infrastructure and services provided by governments. Many government development budgets have tended in the past to neglect agriculture in favour of industry. More recently there has been a widespread tendency to increase the emphasis on agriculture.

To satisfy the increase in world demand generated by even a medium level of population and income growth, the world's agriculture must provide, during each decade remaining in this century, an additional annual output of about 200 million tonnes of cereals, 30-40 million tonnes of sugar, about 100 million tonnes of vegetables, and nearly the same amount of fruits per year; plus about 50 million tonnes of meat and 100 million tonnes of milk – together with the increased livestock feed to produce these products. It is therefore quite apparent that to meet food production needs, particularly in developing countries, will require great efforts both by individual countries and through increased international efforts to assist them to modernize their agriculture.

We know that it is technically possible to increase food production to accommodate the needs of a growing world population for some years ahead with current technology. The new technological packages that are being introduced as a result of increased agricultural research by national and international research centres have the potential for greatly increasing yield. So it is not technology that is the major limiting factor, but the social and economic constraints which retard the wide use of that technology. These problems must be solved in the developing countries, because if the industrial countries could increase their agricultural production substantially to meet the increasing needs of the developing countries it would probably be impossible to find a means of financing and transporting this volume of food.

## Agricultural change in the developing world

Although it might be feasible to supply the increasing food needs of developing countries through stepped-up production in the highly developed agricultural countries, the idea has been generally discarded because of the almost insurmountable problems involved. The monetary requirement within a decade would become so enormous as to appear insoluble for both the exporters and the recipients. The general inadequacy of the developing countries' ports, storage facilities, internal transport, and general infrastructure to handle the volume of imports necessary means that a staggering capital investment and organized effort would be needed in a short time. A rapid acceleration of agricultural development in the food deficit countries therefore appears to be the best solution to the inevitable need for greatly increased food supplies.

This is not a sudden realization, nor has the world agricultural community been caught totally unprepared. There has been a steadily growing and increasingly coordinated effort by the various international organizations, the donor countries, and the private sector to help the developing countries in their efforts to improve and modernize agriculture. An example of this has been the coordinated development of a world-wide network of international agricultural research centres of the highest quality, financed through the Consultative Group on International Agricultural Research, which is sponsored by FAO, UNDP, and the World Bank. Since the World Food Conference, similar new coordinated efforts have been initiated to cover tne non-research needs, such as investment and technical assistance.

It is not possible simply to transplant the technology of the highly developed agriculture of the industrial countries to fill the needs of the developing world. In many cases this has been tried and an unhappy lesson learned through expensive failures. There are various reasons why the requirements tend to be very different in developing countries. Some of these are due to cultural, social, economic, and political differences; others are the result of geography. Nearly all the industrial countries' agriculture is carried out in temperate climates. In contrast, the greatest concentrations of population and agricultural land in the developing countries are in tropical or sub-tropical areas. The agricultural practices — in fact to a large extent even the primary crops — tend to be totally different in the industrial and the developing countries.

However, there is one major similarity: the need for increased productivity per unit area through intensification of agriculture. Although there is still a substantial amount of potentially useful land that could be developed for agriculture, particularly in parts of Africa and South America, most of this could be developed only at a high capital cost. There is also a serious question about the suitability or advisability of diverting much of this land to agriculture; already there exists agricultural use of substantial areas of highly erodible land which should be returned to forest or similar conservation uses.

The year-round growing season in tropical areas facilitates a high degree of intensification of production through multiple cropping and other methods, if adequate water, fertility, and pest control can be provided. But under intensive agriculture, pest problems become more severe; and in the tropics, where no seasonal break is provided by winter to arrest the build-up of pest populations, a crop can be completely lost through attack by pests. The favourable climates for pests in the tropics have also led to a much greater diversity of pest problems in these areas.

### Pesticides in food production

Although various methods are used to control pests in different pest management systems, pesticides are at present the most important factor in most control programmes, and this will be so for the foreseeable future. In 1974, the world pesticide market was about 5000 million U.S. dollars, of which about 40 per cent was in North America and only slightly less in Europe. The developing countries accounted for no more than 10 per cent of the total. The rate of increase in the use of pesticides in developing countries is considerably higher than in the intensive agricultural countries, and this trend is expected to continue.

Not only are pesticides the primary immediate weapon against pest losses, both on growing crops and during transport and storage after harvest; their use in chemical weed control is also a major factor in increased labour efficiency and reduced drudgery. The herbicides used for chemical weeding now represent more than half of all pesticides used. Adequate use of pesticides has a major stabilizing effect on agricultural production by preventing periodic massive losses caused by insect and disease outbreaks. Consistent production is very important for price stability.

Adequate and proper use of pesticides can be a major factor in achieving a better environment through improved public health resulting from control of fly, mosquito, and other vector-borne diseases; better health through adequate nutrition; supplies of reasonably priced, wholesome food; and less pressure on the world's limited and precious land resources by making it possible to produce more food on less land. The proper use of pesticides would thus make it possible to confine agriculture to the most productive and agriculturally suitable land, and so eliminate the need for further destruction of forest to provide more arable land. The more erodible hill and other land could be kept out of agriculture, and marginal land of this type which is now cultivated could be returned to pasture, forest, or other conservation uses including recreation and reserves for wild life.

The pesticide industry is one of very high technology and is research-intensive. This has to a large degree limited the primary pesticide manufacturing industry to the highly developed industrial countries. Public concern about the potential hazards that some people feel pesticides could exert on the environment has resulted in the pesticide industry being one of the most stringently regulated industries in the world.

Not only is there the need to increase world pesticide production to ensure an adequate food supply; there is also the need for continued high research investment to discover new, more efficient, and safer pesticides, and to find better and safer ways of using those which are currently available. The increasing array of divergent and more stringent world regulatory requirements (see Chapter 19) threatens to take the time and investment required to develop new pesticides beyond the point that will justify the investment. Indeed, it is ironic that the research requirements placed on pesticides to ensure environmental safety have already often made it commercially unattractive to develop those highly specific pesticides which provide the greatest degree of environmental safety. Because of their very specificity, the market is limited and the cost-risk-benefit ratio does not justify investment.

The safety record in pesticides is, perhaps, nearly the best in any area of modern technology. Their benefit to mankind has been one of the greatest through their contribution to improved health and economical food production. They are urgently needed in the future to help ensure adequate food for survival

Much of the anxiety about pesticides is derived from people who have unfounded fear about all synthetic chemicals who seek to consume or come into contact only with 'natural' substances. They should realize that nature is the most skilled organic synthesizing chemist known, and that man has so far been able to duplicate only some of the simpler natural organic compounds. Many of the natural compounds synthesized by plants and animals are highly toxic or persistent in the environment, or both; and many everyday foods contain natural toxins.

There needs to be an increased public awareness of the facts about pesticides: an awareness that will prevent fear and hysteria emanating from a few over-zealous and misguided individuals from leading to a global tragedy that will entail mass suffering or starvation for the poorest people and serious harm to all.

The food we eat is limited to that derived from a small percentage of the world's animals and plants, because many of the latter are too toxic for man to consume. Even closely related plants in the same family as some of our common food crops are too poisonous to eat, and some of the most widely consumed tropical tubers must be soaked or cooked to destroy toxic substances before they are edible. Few people realize that potatoes and tomatoes, which can be eaten because they do not contain excessive levels of natural toxins, are close relatives of the nightshades, which are very poisonous, The same is true of the lettuce group, and also that group of closely related plants which include carrots, celery, parsley, and parsnips; these we can eat, but we cannot eat other plants in the group, such as the hemlocks: most schoolchildren know that Socrates committed suicide by consuming a tiny quantity of poisonous hemlock.

Many of the same factors are true of natural insecticides. A number of today's household sprays used to kill flies, mosquitoes, and similar insects contain a natural insecticide extracted from the pyrethrum plant. In the past, nicotine was widely used as an insecticide until it was replaced by safer and more efficient synthetic compounds. It is doubtful whether a natural compound like nicotine could qualify under the current stringent regulations for pesticides, for it is known to be toxic. With these thoughts in mind, it must be clear that a continuing supply of pesticides is a basic factor in our future welfare.

# 2 Population and food production

by R.H. Gray

In all animal species, the size and density of a population is in part determined by the availability of food supplies. Man is no exception to this rule, and since *Homo sapiens* evolved about 2½ million years ago, the human population has had to maintain a balance between numbers and resources. Thus, the historical growth of the human population is closely linked to changes in the supply of foodstuffs as determined both by natural factors and by the technological innovations which altered man's productive capacity.

Populations grow when there are more births than deaths, and throughout most of man's history high death-rates have acted as the main check on population growth. This high mortality was a product of the three great scourges of mankind: famine, disease, and war; and of these three, famine or the availability of food supplies was by far the most important factor limiting population increase.

Thomas Malthus, writing in 1798, was the first to describe the relationship between demography and ecology. He argued that the productivity of the land was the ultimate basis for human survival and that 'the power of population growth is infinitely greater than the power of the earth to produce subsistence for man'. Malthus concluded that population growth could not continue indefinitely, and that it must inevitably slow down or cease owing to the action of 'positive checks' such as starvation, which increased mortality, or 'preventive checks' such as delayed marriage, which would reduce fertility. However, Malthus was the child of a traditional agricultural society, and although his analysis was essentially true of the pre-industrial era, he failed to appreciate the potential impact of the nascent Industrial Revolution, which dramatically altered man's productivity and his relation to the environment.

Before the agricultural revolution which occurred about 10 000 years ago, man was a hunter-gatherer living at bare subsistence levels. The ecological balance between population and food supply was precarious and the density of human settlement was probably less than one person per square kilometre. It is impossible to estimate the

13

Plate 2.1. Most of the prospective world population growth will occur in the developing countries, where in-creases of 78 per cent are expected between 1975 and 2000. This will call for a doubling of food production, involving much greater use of fertilizers and pesticides. Today 70 per cent of the population lives in countries which provide family planning services: this New Delhi bus stop carries propaganda in English and Hindi. (FAO photograph by P. Pittet)

size of the population during this remote prehistoric period, but archaeological evidence suggests that human societies consisted of small scattered communities and that total numbers were restricted by available resources. Furthermore, fossil remains indicate that the expectation of life rarely exceeded 20 years, and such a short life-span is incompatible with any significant population increase.

The development of agriculture and the domestication of animals revolutionized the balance between food supplies and population. Settled communities emerged and generated a small agricultural surplus which could sustain the growth of limited urban centres and allow economic specialization in non-agricultural activities. These advances, allied with the discovery of storable grains, led to the growth of great civilizations based on staple foodstuffs, and the era of settled agriculture continued with relatively little modification until the Industrial Revolution of the eighteenth century.

The agricultural era was not a period of demographic stability. During good years, when births exceeded deaths, the population often grew at a modest rate. This growth, however, was periodically halted or even reversed by major increases in mortality. The picture that emerges from historical demographic studies is one of slow and erratic population growth punctuated by violent fluctuations in the death-rates. These fluctuations in mortality were often closely associated with famine, and detailed studies of European communities during the sixteenth and seventeenth centuries demonstrate a clear correlation between bad harvests and increases in the death-rate.

The expectation of life during the agricultural era was very variable. In the city of Rome, for example, it was about 20 years, whereas in the more salubrious Roman colonies of North Africa, it exceeded 40 years. In England during the Middle Ages, expectation fluctuated from some 35 years during the thirteenth century to 17.3 years during the Black Death in 1348. Even in the eighteenth century, people in Europe rarely lived beyond 40 years.

Given this unstable situation, it is difficult to estimate the size of the world's population, but it probably increased from around 5 to 10 million at the beginning of the Christian era, and reached approximately 725 million by the mid-eighteenth century. The average rate of population growth was less than 0.15 per cent per annum over this long period, and at such a slow rate it would have taken almost 500 years for the population to double. By the year 1800, however, the world's population had increased to about 900 million; in 1900 it was about

1500 million; and by 1975 it had almost reached 4000 million (Fig. 2.1)
At the current rate of growth of 1.93 per cent per annum, the world's
population will double within 35 years.

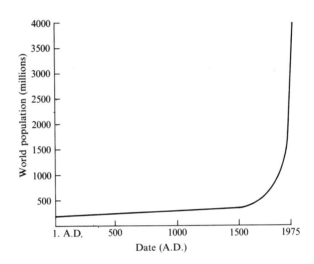

Fig. 2.1. The growth of world population from A.D. 1 to 1975.

The rapid and dramatic increase in population over the past 250
years resulted from an unprecedented decline of mortality which
occurred first in the now industrialized countries, and only much later
in the currently developing nations. To understand these demographic
processes, therefore, it is necessary to consider the changes separately
for the developed and developing regions of the world.

### The demographic transition

The history of human population growth can be divided into a number
of phases known as the 'demographic transition'. In its simplest form,
the demographic transition consists of a series of steps whereby
societies move from a state of high birth-rates and death-rates to a state
of low birth-rates and death-rates. This is essentially a descriptive
model, and not all countries have followed the same demographic path,
but there are enough similarities in the historical experience for this
model to be a useful theoretical device.

The first, pre-modern phase of the transition consisted of a situation of high and fluctuating mortality associated with high fertility and early marriage. During this period births and deaths were almost in equilibrium and the rate of population growth was slow or negligible. However, in some pre-modern European societies the age of marriage was delayed, and although no attempt was made to control fertility within marriage, the late marital age tended further to reduce population growth.

The second major phase of the transition consisted of a decline in the death-rate with continuing high fertility and consequent rapid population growth. The third and final phase of the transition is characterized by death-rates and birth-rates that are in equilibrium at a low level. Fertility is generally controlled by family planning and marriage is generally delayed to a mature age.

The industrialized countries have experienced all three phases of the demographic transition, whereas many developing countries are now only in the second phase and some of the poorest nations have barely emerged from the first phase. One cannot, however, extrapolate from the historical experience of the industrialized countries in order to predict the future course of population growth in the developing countries: there are profound differences in the social and economic processes underlying the demographic transition in the developed and the developing worlds.

## The demographic transition in the industrialized countries

During the past two centuries the populations of the industrialized countries have increased from approximately 201 million in 1750 to 834 million in 1950, and by 1975 the developed world contained 1133 million people. This vast population growth was unprecedented in all of history, but even during the period of maximum increase, the rate of population growth seldom exceeded 1.5 per cent per annum, and some of the surplus population in Europe was siphoned off by migration to the New World.

The decline of mortality, or second phase of the demographic transition, began first in western Europe during the mid-eighteenth century, and spread throughout the developed nations over the next 150 years. The timing and rate of fall of mortality varied considerably from country to country, but there are basic similarities in the demographic pattern. The beginning of the second phase of the demographic transition was characterized by the elimination of major fluctuations in mortality and the stabilization of death-rates at a relatively high level. Subsequently, the stable death-rates decreased slowly and steadily until after the Second World War, when they reached current levels.

The birth-rates generally remained high during the second phase of the transition, although the age of marriage was generally delayed and a significant proportion of the population in certain countries remained unmarried. Married couples tended to have very large families, and there was a considerable time-lag before the onset of the third transition phase when fertility control became widespread throughout the industrialized world. Even in recent decades, the birth-rates in the industrialized countries have fluctuated considerably in response to social and economic conditions. For example, birth-rates fell to a very low level during the Depression of the 1930s and during the Second World War, but this was followed by a post-war 'baby-boom' which continued until the mid-1960s.

The fall of mortality which initiated and sustained the demographic transition was due to a reduction of deaths from infectious diseases and malnutrition, especially among infants and children. These changes were closely linked to the spread of the Industrial Revolution, which brought about improved standards of living, better sanitation and hygiene, and most important, a significant improvement in general nutrition. Advances in medicine, however, had little impact on overall mortality until the late nineteenth and early twentieth centuries.

Several factors contributed to improved nutrition during the second phase of the demographic transition in the developed world. Agricultural productivity had been increased by the introduction of new staple crops such as the potato, better land management, new systems of crop rotation, the improvement of seed varieties, and later the mechanization of agriculture. The increasing use of mechanization in agriculture decreased the requirement for rural labour, and produced a surplus work-force which moved from the rural areas to the newly developed industrial towns and cities. In Britain the proportion of the population employed in agriculture decreased from some 40 per cent at the beginning of the nineteenth century to 20 per cent about 1850 and 9 per cent by 1900; and by 1850 half the British population lived in urban areas. The growth of agricultural output created both the food supplies required to sustain a large urban population and the work-force to man the new industry.

The improvements in agricultural productivity were augmented by new developments in transport and communication. Road networks were extended, and the construction of canals and railways increased the transport of food from rural areas to the growing urban centres. The shipping industry expanded to facilitate the importation of cheap surplus grain from the new colonial wheatlands of North America and Australia, and countries such as Britain became progressively dependent upon imported foodstuffs. Furthermore, during the latter half of the nineteenth century the development of the canning and refrigeration industries increased the shipment of meat into Europe.

The economic prosperity associated with industrialization enhanced the standard of living. Not only was more food available, but the average man was able to afford a better diet. The growing purchasing power of the urban-industrial population acted as an economic stimulus to the development of agriculture and the food industry. Changes in food production, distribution, and storage reduced the threat of harvest failure and improved the seasonal variety and quality of food. Thus the Malthusian prediction of mass starvation was averted by technological innovation.

## The demographic transition in the developing world

The demographic experience of the developing countries has been profoundly different from that of the industrialized nations. The developing regions of the world contained approximately 920 million people in 1850 and the population grew only marginally to 1100 million by 1900. However, the population had increased to 1683 million in 1950, and by 1975 it had almost doubled to over 2800 million. The growth of population in the developing countries has clearly been a recent phenomenon; in these countries the annual rate of population increase has accelerated over the past 25 years from 0.5 per cent to about 2.4 per cent in 1975. This rate of increase is much faster than the 1.5 per cent peak rate of population growth experienced by the industrialized countries; and, unlike the European experience, there is little possibility of migration easing the population pressure.

The excessive population growth in the developing world is due to two factors: the rapid decline of mortality; and the persistent high level of fertility, which is significantly greater than that of the industrialized countries during the nineteenth century. The high fertility in the developing world results from a social pattern of early marriage and the absence of widespread birth control.

The decline of mortality in the developing countries has been extremely variable, and there are still great contrasts in death-rates.

Many countries in Africa have an expectation of life below 40 years, whereas some of the richer countries in Latin America and Asia have an expectation of life approaching the European levels of 70 years. In some countries the death-rates began to fall slowly during the first quarter of the twentieth century, but in most of the developing nations significant declines of death-rate did not occur until after the Second World War. The reduction of death rates has, however, frequently been very rapid: in Britain it took 150 years for the death rate to fall from 25 to 12 per thousand, whereas this transition was achieved in little more than a decade in countries such as Mauritius and Ceylon.

As was the case in the industrialized countries, the fall of mortality was largely due to a reduction in deaths from infectious diseases, especially among infants and children. There are, however, marked contrasts between the mechanisms underlying the decline of infectious disease mortality in the industrialized world during the nineteenth century and the declines in the developing countries over the past three decades. In the industrialized countries, infectious diseases were largely controlled by improved standards of living, nutrition, sanitation, and hygiene. But in the developing countries medical technology and major public health programmes have been the most important factors.

The period following the Second World War was a watershed in the history of medical science. During this era a wide variety of new and effective drugs were developed. Antibiotics to combat bacterial infections and drugs against malaria and other parasitical diseases became widely available. Residual insecticides such as DDT facilitated the control of major insect-borne diseases and led to world-wide campaigns against malaria. New vaccines provided protection from a broad spectrum of infections, and techniques of production and storage of vaccines led to mass immunization campaigns in many countries. This new technology was effective, relatively cheap, and simple to use. For example, one poorly trained man equipped with a DDT spray could save more lives from malaria than a highly trained doctor.

During the post-war years new administrative systems emerged to enhance the application of this technology. The developing countries took major strides in expanding the health-service coverage, especially for the poorer groups in the population. International bodies such as the World Health Organization helped to coordinate campaigns against a number of major diseases and helped to transfer knowledge and technology to the developing world. Furthermore, during the period of decolonization and the post-colonial era, the rich industrialized countries have provided aid for health programmes in the poorer nations.

Socio-economic development and the improvement of living standards contributed to the decline of mortality in many developing countries. Their historical experience was, however, different from that of the industrialized world, for socio-economic development has ceased to be a prerequisite for reducing mortality, since even in very poor underdeveloped areas death-rates can be reduced by medical technology. Thus, the fall in mortality in the developing world has, in part, been dissociated from the general socio-economic improve: ment.

There is one further crucial difference in the experience of the rich and the poor worlds. The growth of population in the industrialized countries was stimulated and accompanied by a more than commensurate increase in food production associated with improved agriculture and the opening up of new land. In many developing countries techniques of agriculture are, however, still traditional and relatively inefficient; and there is little good land left for future exploitation – arable or grazing. In Bangladesh the population density is already 840 people per square kilometre of arable land, and there is no potential for expanding the area under cultivation. Food production, therefore, is perhaps the most important factor determining the future population growth in the developing countries.

## Population growth and food production, 1950-75

During the 1950s and 1960s the increase in world agricultural output kept pace with the total growth of population, and there was a net increase of food production per head. However, rapid population growth in the less developed countries eroded the agricultural gains so that the *per caput* increase in food production of the poorer nations was significantly lower than that of the richer countries (see Table 2.1). In one-third of the developing nations, population growth outpaced food production during the two decades from 1950 to 1970.

More recently, a series of bad harvests and droughts in the developing countries has adversely affected agricultural development, and it is estimated that food production rose by only 1-2 per cent in 1971 and 1972, and by less than 1 per cent in 1973. This growth of food production was less than the annual rate of population increase of 2.4 per cent. Thus, during the 1970s food production per head in the developing countries has actually declined, and the poorer nations have

Table 2.1. *Rates of growth of population and food production 1952-72*

| | Growth rates (per cent per year) | | | | | |
| | 1952-62 | | | 1962-72 | | |
| | Population growth rate | Total food production growth rate | Food production per caput growth rate | Population growth rate | Total food production growth rate | Food production per caput growth rate |
|---|---|---|---|---|---|---|
| World | 2.0 | 3.1 | 1.1 | 1.9 | 2.7 | 0.8 |
| Developed regions | 1.3 | 3.1 | 1.8 | 1.0 | 2.7 | 1.7 |
| Developing regions | 2.4 | 3.1 | 0.7 | 2.4 | 2.7 | 0.3 |

Adapted from ref. (3), p.5.

been forced to import ever-increasing quantities of grain from the developed world. The price of these imports has escalated with rises in production costs and the pressures of demand on falling world food reserves.

These global figures of food production do not fully reveal the problems facing the Third World, since inequalities of income severely limit the purchasing power of the poor, who often cannot afford to buy available food. Surveys suggest that approximately 40 per cent of children in developing countries suffer from various degrees of malnutrition, and that a total of 400 million people have grossly inadequate diets. FAO has estimated that food supplies must expand by 4 per cent per annum in order to keep pace with population growth and to compensate eventually for existing deficiencies, but recent productivity has fallen far below this target.

Food production might be increased either by expanding the land under cultivation or by increasing the yield per hectare. It is difficult to estimate how much currently unused land could be brought under cultivation, for the limiting factors are largely economic and socio-political constraints. FAO has calculated that only 56 per cent of potentially arable land is now under cultivation, though the scope for future expansion varies from zero in the densely populated countries of Asia to as much as 68 per cent in Africa and 79 per cent in Latin America. An extension of the cultivated area, however, would require a huge capital investment of around $500 to $1000 per hectare; this is far beyond the capacity of most developing countries.

Given the constraints on the expansion of cultivation, the most promising alternative is to increase productivity; here there is considerable scope for improvement. Traditional farming practices are often inefficient. For example, rice yields in Bangladesh and India are only one-third of those obtained in Japan, and corn (maize) yields in countries such as Thailand and Brazil are less than one-third of those achieved in the USA. Therefore, the modernization of agriculture by the introduction of improved crop varieties, the extension of irrigation, and the wider application of fertilizers and pesticides offer hope for significant increases in productivity. Such increases are vital if we are to overcome the present deficiencies, and, more important, if we are to cope with future population growth.

## Projections of future population growth and food requirements

Demographic projection is fraught with uncertainty, but it is possible to make reasonable predictions of the likely course for future population growth and to estimate the maximum population size at various times in the future. Population projections are usually made over a medium term of 25 to 30 years. Such medium-term projections are based on a set of assumptions about present-day trends, and they give a reasonable picture of the demographic prospects for the next generation. However, the process of population growth has an inbuilt momentum which carries on beyond the length of one generation. This momentum is created by the existing age structure of the population, for today's children are tomorrow's parents, and already the future parental generation is numerically much larger than the present parental cohort. In most developing countries 40–5 per cent of the people are under the age of 15, compared to 20-30 per cent in the industrialized world, and these large groups of young people will reach the child-bearing ages over the next two to three decades. It is therefore necessary to undertake long-term projections to account for the effects of the existing age structure, and to assess the prospect for eventual stabilization of the population.

The United Nations has prepared a series of medium-term projections up to the year 2000. According to these estimates, the total world population is likely to grow from 4000 million today to approximately 6500 million by the year 2000 (Fig. 2.2). If birth-rates do not decline significantly over the next 25 years, the population could reach 7200 million. Alternatively, if the decline of fertility is extremely rapid, the population may reach only 5950 million. Most of this prospective population growth will occur in the less developed countries, where the populations are likely to increase by 78 per cent from 2800 million in 1975 to about 5000 million at the end of the century. In contrast, the populations of the industrialized world are likely to grow by 28 per cent from 1133 million today to 1450 million in 25 years' time. Thus, short of a major catastrophe, the medium-term demographic history of the world will be largely determined by events in the developing countries.

To cope with this increase in numbers over the next 25 years, total food production will have almost to double by the end of the century. The major increase in production must occur in the developing countries so as to keep pace with the population growth, to compensate for existing shortfalls, and to avoid the enormous economic burden of increased food imports. A doubling of world food production would require a more than three-fold increment in the annual use of fertilizers,

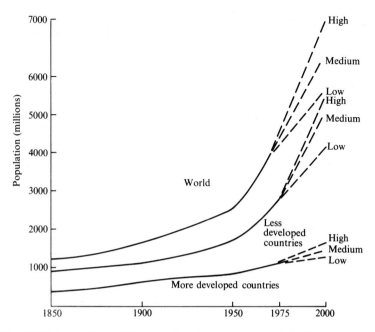

Fig. 2.2. Total world population growth and population growth in the more and less developed regions 1850-1975, with projections to the year 2000 at three levels of fertility (high, medium, and low).

and much more extensive use of pesticides. The energy costs of these chemicals and of agricultural modernization would, however, be large, since in non-industrialized countries up to one-third of the total energy use may have to be devoted to agricultural production.

Looking further ahead, demographers have prepared projections to estimate the population size around the year 2100. If there is a great decline of fertility by 1985, the world's population could stabilize at about 6400 million people in 2100. Such a course of events is unlikely, and it is more probable that the population will reach 11 200 to 15 000 million by the end of the next century. Even then, population growth would probably not cease before the year 2150. The less developed countries will bear the brunt of this future increase and these countries will eventually contain between 80 and 88 per cent of the world's total population.

It is impossible to predict whether food production can be increased sufficiently to sustain such large populations. Theoretical estimates suggest that it would be possible adequately to feed up to 38 000 million people, but this would require radical changes in technology, a massive transfer of resources from the rich to the poor world, and an unprecedented investment in agriculture. Therefore, in the long term, political and economic factors will determine whether food supplies can keep pace with population growth.

## Family planning and the prospects for fertility reduction

The demographic transition in the industrialized world involved a decline in death-rates followed by a decline in birth-rates. Most developing countries have experienced significant falls in death-rates but in many the birth-rates have remained high and the future course of population growth in these countries will be largely determined by the future pattern of decline in fertility.

There are two broad preconditions for a reduction of fertility. First, parents must see fertility control as advantageous and acceptable; second, techniques for the control of fertility must be widely available and widely used. At present, neither of these preconditions is fully realized in most developing countries. Large families are still perceived as an economic advantage because children provide productive labour and security for their parents' old age. The fear of a child's death leads to a desire for more children as a form of 'insurance', and in many traditional cultures a woman's social standing depends on the number of children she has borne.

In addition, there are often religious or social obstacles to the adoption of modern contraception. Modernization, socio-economic development, and an increase in the standard of living will probably alter this situation, but without such progress the future for fertility control may be bleak. On the other hand, the use of modern contraception is increasing throughout the world. Experience over the past decade suggests that birth control can contribute to significant falls of fertility if the services are sufficient to reach the bulk of the population, and if the political, social, and economic conditions are favourable. The existence of contraception cannot by itself reduce fertility; it can only provide the means for individual couples to control their fertility if they so wish.

In countries such as China, Taiwan, Korea, Singapore, Hong Kong, Mauritius, and some Caribbean islands, family planning has met with considerable success. Equally, in large countries such as India, Pakistan, Bangladesh, and Kenya, the programmes have had only limited impact; and there are still many developing countries in Africa and Latin

America where birth control is either not available or is only provided on a very limited scale. However, 70 per cent of the world's population now lives in countries which offer family planning through existing public health services, and the programmes are growing rapidly. There are, therefore, reasonable grounds for some optimism; but one must accept that even if the preconditions for fertility decline exist, and even if birth control is widely adopted, the world's population will still continue to expand until well into the twenty-first century. It follows that food production must be increased to meet the demands of inevitable population growth.

### Further reading

1. ALLISON, A. *Population control.* Penguin Books (1970).
2. CIPOLLA, C.M. *The economic history of world population.* Penguin Books (1967).
3. *Population, food supply and agricultural development. World Population Conference, Bucharest, August 1974.* Conference background paper, E/CONF.60/CBP/25.
4. FREJKA, T. The prospects for a stationary world population. *Scient. American.* **228** (3) 15 (1975).
5. United Nations Organization. *World population prospects as assessed in 1968.* UN Publication No.E.72.XIII. 4, United Nations Organization, New York (1970).
6. ——, *The determinants and consequences of population trends.* UN Publications, Population Studies No. 50, Vol. 1. United Nations, New York (1973).
7. ——; *Concise report on the world population situation in 1970–75 and its long-range implications.* UN Publications, Population Studies No. 56. United Nations, New York (1974).
8. WOLFERS, D. Problems of expanding populations. *Nature, Lond.* **225**, 593 (1970).
9. WRIGLEY, E.A. *Population and history.* Weidenfeld and Nicolson, London (1969).

Plate 3.1. An African mutilated with elephantiasis caused by the mosquito-borne disease filariasis. In some tropical areas up to 30 per cent of the population may suffer from elephantiasis swellings. Control of the disease is by spraying the mosquito larvae with certain organophosphate insecticides, and by mass administration of drugs such as diethylcarb-amazide. (Shell photograph)

Plate 3.2. No one can escape contact with water, and most of the tropical vector-borne diseases are related to water. Often the prevalence of such diseases as schistosomiasis, malaria, and onchocerciasis has been increased by the building of man-made lakes and the introduction of agricultural irrigation schemes in the tropics. (Shell photograph)

# 3 Vector-borne diseases and the need to control them

G. Davidson

Some of the most common diseases in the world are passed from man to man by invertebrate vectors. Most of these carriers are insects, ticks, and mites, but in an important group of diseases snails are involved. Most vector-borne diseases are tropical and are most prevalent in developing countries. Those, such as the louse-borne diseases, that occur in the colder parts of the world are invariably associated with poor standards of hygiene which may be the product of underdevelopment or associated with wars and famine.

For a number of reasons it is extremely difficult to assess the importance of these diseases in strictly economic terms. Records of morbidity and mortality attributable to specific diseases, or even reliable records of births, deaths, and population numbers, do not exist for most developing countries. In fact, in most cases the diseases take their toll without any contact with an organized health service of any sort. According to the World Health Organization,[12] reliable records of vital events exist for only about one-third of the world's population, and probably only about 2 per cent of infant deaths are notified. According to this source, the expectation of life at birth is the best single indicator of the health status of a population, and a strong correlation exists between it and the *per capita* gross national product. In the developing regions, where three-quarters of the world's population lives, this expectation has been estimated as 50 years, as compared with 70 years in the developed countries. In all probability most of this difference can be accounted for by vector-borne diseases.

It is extraordinarily difficult to produce reliable figures on the incidence of disease in developing countries and the resulting mortality. A large scale attempt to do so from 1957 data from 169 countries and territories with a total population of 1 204 000 was made by Wright *et al.* in 1962.[16] This showed glaring anomalies and it seems more than likely that emphasis has been placed on diseases exotic to the reporting country to the neglect of the indigenous and commonest diseases. Ansari and Junker[1] have presented these data on the five principal

vector-borne diseases of the world in terms of prevalences in those parts
of the populations which actually reported the diseases. These data all
set out in Table 3.1. Proportions of populations notifying diseases varied
from as little as 0.82 per cent for filariasis in Middle Eastern countries
to as high as 94.15 per cent for malaria in the Caribbean and Central
and South America. Table 3.1 further transforms the data for the five
main vector-borne diseases into actual numbers of cases and actual
numbers of deaths. The results produce what are now considered to be
gross underestimates of the actual situation and it would appear that
the analysis serves only to indicate the relative prevalence of the five
diseases.

### Malaria

Malaria is the most common of all the vector-borne diseases, in spite of
the numerous attempts that have been made to eradicate it since the
1950s. Eradication is now considered a practical impossibility in most
of the centres of distribution of this disease. In September 1974, of the
1945 million people living in originally malarious areas, 1422 million
(or 73.1 per cent) were in areas where malaria is eradicated or where
attempts are now being made to eradicate it. The remaining 523 million
(26.9 per cent) live in areas where there is no eradication programme.[14]
According to Pampana and Russell,[10] the annual incidence of the
disease before eradication was about 250 million cases, of whom some
2.5 million died. The present annual incidence is of the order of 100
million. As has been pointed out by Bruce-Chwatt, [2] eradication
programmes almost certainly prevented the occurrence of more than
2000 million cases of malaria over the years 1961-1971. Bruce-Chwatt
also estimates that 15 million people were saved from dying between
1955 and 1965, during which time the world's population increased by
500 millions: this puts in perspective the relative effect of malaria
eradication on population increase.

Broadly speaking, although it involves four different species of
blood parasite, malaria is of two main epidemiological types. First,
there is a stable type resulting from continuous transmission by a long-
lived, predominantly man-biting, anopheline mosquito. The second type
is an unstable or epidemic variety resulting from seasonal transmission
by a short-lived anopheline mosquito that is more catholic in its feeding
habits. The first type produces a high degree of immunity in individuals
above the age of about 10 years, when the disease has little
apparent effect on general health; but below this age, children suffer

from intermittent fever and an enlarged spleen, and a significant percentage of them die before reaching the age of five. According to Lepeš,[7] one million children die from malaria every year in tropical Africa. This is the form the disease usually takes in tropical Africa and

Table 3.1. *The five principal vector-borne diseases of the world*

| Disease | Infectious Agent | Vector | Distribution | Number of cases* | Number of deaths* |
|---|---|---|---|---|---|
| Malaria | Protozoan | Mosquito | World-wide, tropical and sub-tropical | 4 155 525 | 313 170 |
| Filariasis | Helminth | Mosquito | World-wide, tropical | 264 870 | 53 |
| Schistosomiasis | Helminth | Snail | Middle East, Africa, Central and South America, South-east Asia | 260 172 | 530 |
| Trypanosomiasis | Protozoan | Tsetse fly Triatomid bug | Africa Central and South America | 11 443 | 1180 |
| Onchocerciasis | Helminth | Blackfly | Africa, Central and South America | 7588 | 8 |

*Derived from morbidity and mortality rates given in ref. (1) and based mainly on WHO and FAO figures for 1957 taken from ref (16).

because it has little effect on the adult population its importance has been underrated.

The second variety of malaria produces little immunity, intermittent illness (sometimes severe) in all ages, and a continuously enlarged spleen. This is the form that is characteristic of the subcontinent of India with its mammoth-scale epidemics. The latest and most spectacular example is the occurrence between 1968 and 1970 of some 1.5 million cases (out of a total population of 12 million) in Sri Lanka after eradication procedures had reduced the number of infected people down to 18 in 1963. Fortunately, mortality from this epidemic was low; far less than the 80,000 deaths that resulted from the 1934-5 epidemic in the same country.[3]

By far the greater portion of malaria eradication has been achieved by the use of residual insecticides applied as surface deposits to the internal walls and roofs of human habitations. According to Bruce-Chwatt,[2] 40000 tons of DDT were used annually in public health programmes (mainly against malaria) during the 15 years from 1957 to 1972. During this period no major mishap attributable to the toxicity of DDT was recorded among the 200000 men who sprayed this insecticide. This represents only 15 per cent of the world's DDT production and as it is applied largely indoors it produces a minimum of environmental contamination.

Insecticide resistance has so far been detected in some 50 species of anopheline mosquitoes, some 20 of them important vectors, but few campaigns have been brought to a complete halt by it. (The whole question of resistance is discussed in Chapter 16.)

### Filariasis

Filariasis is the disease which produces the most spectacular of all the mutilations that man is subject to, elephantiasis, in which various parts of the body swell to enormous proportions. In areas of high transmission as much as 30 per cent of the population may be affected by such swellings.

In 1974 WHO estimated that 250 million people were infected with *Wuchereria bancrofti* and *Brugia malayi*, the causative agents of filariasis, although new, more sensitive, diagnostic techniques could easily double this figure.[13] This represents the total number of cases in existence, not the annual incidence, and it should be remembered that like all the other worm diseases the parasite may be present in the body for many years. Various mosquito species (both anopheline and culicine) carry the disease, which can be both rural and urban. In fact, with increasing urbanization in developing countries, where over-crowding and inefficient sanitation often go hand in hand, a large increase in the urban mosquito vector, *Culex pipiens fatigans,* has taken place. Tropical Africa, south-east Asia, and many of the Pacific Islands are the areas of greatest incidence. In India alone 122 million people live in areas where Bancroftian filariasis is endemic. The severity of the disease is very much a function of the frequency with which the individual is reinfected, and of the general 'worm-load' of the body. Low-intensity infections of microfilariae seem to have little effect on general health. In India, some five million people are estimated to suffer from the disease, while up to 13 million have microfilariae circulating in their blood.

In more than one instance, house-spraying with insecticides, though nearly always primarily directed towards the control of malaria, has also led to a reduction in anopheline-transmitted filariasis. However, insecticidal control of the urban factor, *C. p. fatigans,* presents problems in that the natural tolerance level of this species, especially in the adult stage, is usually much higher, and resistance to the organochlorine insecticides is common. Larval control is usually preferred with such insecticides as the organophosphates fenthion, Dursban, and chlorfen-vinphos. These are particularly suitable for the water of high organic content favoured by this species. Experimental work is now going on in India to investigate the potentiality of *C. p. fatigans* control by genetic means. These involve large-scale production of sterile or incompatible males for mass release into wild populations. Success in this direction would obviate the use of insecticides. (See also Chapter 20.)

Filariasis is one of the few vector-borne diseases to be successfully controlled by mass drug administration. Diethylcarbamazine has produced spectacular decreases in *W. bancrofti* in Polynesia, although it must be admitted that some of the resulting low parasite counts still remain infective to mosquitos.[11] Such success has not been achieved in India.

### Schistosomiasis

The symptoms of schistosomiasis vary from mild pains in the bladder, kidneys, and abdomen, headaches, dizziness, and diarrhoea, to the chronic and severe illness resulting from permanent damage to the kidneys, liver, bladder, and intestines, and occasionally to the spinal cord and brain.

Three principal species of trematode worms of the genus *Schistosoma* are responsible for the disease we know as schistosomiasis. They are *S. haematobium*, which affects the urinary system and is confined to the African continent and the Middle East; *S. mansoni*, which affects the alimentary system and is found not only in Africa but also in South America and the Caribbean; and *S. japonicum*, which also affects the bowel, and occurs in China, the Philippines, Indonesia, and Japan.

As with filariasis, the severity of the symptoms depends to some extent on the 'worm-load'. It is not the adult worms which produce these symptoms (they live for many years inside the veins supplying the gut or bladder), but the eggs, which pass through the walls of the blood vessels and the tissues of the organs they supply before being evacuated in urine or faeces. Some eggs are also carried by the blood-stream to the liver and other organs, producing pathological changes there as well as in the bladder and intestine. Excreted eggs give rise to motile stages which penetrate snail hosts, multiply, and release free-swimming forms which penetrate human skin to start a new infection.

Wright [15] has estimated that 592 million people are at risk from schistosomiasis and 125 million are actually infected. He also estimates that more than 2.5 million of these are totally disabled and nearly 25 million are only capable of 90 per cent productivity as a result of infections of the three species in the Middle East, south-east Asia, and Brazil. The total economic loss is put at $642 m. The work of Forsyth and Bradley [5] in Tanzania indicated that serious damage to the urinary system due to *S. haematobium* was more prevalent in children (20 per cent) than in adults (10 per cent) and might therefore be the cause of mortality in age groups about to begin a productive life.

The greatest hopes for the control of schistosomiasis are placed on molluscicides. These chemicals are capable of killing the snail hosts, especially in areas where breeding places are limited and circumscribed. Drugs are available for treatment of the disease but most have side-effects and can be given only under medical supervision. One of them is an organophosphorus compound and also an insecticide, namely Dipterex or trichlorphon. Fenwick[4] has worked out relative costs of molluscicide and drug treatment for *S. mansoni* infections on a sugar estate in northern Tanzania during a three-year control programme. The molluscicides used were N-tritylmorpholine and niclosamide; the drugs were niridazole and hycanthone. Snail control proved more feasible with a total cost of $23538 to protect nearly 6000 people or $1.31 per person per year. Total expenditure on mass diagnosis and treatment was

$37 043. It was estimated that if freedom from the disease increased
productivity by 5 per cent a saving of some $14 000 a year could result.

### Trypanosomiasis

Trypanosomiasis is the condition resulting from infections with various
species of the flagellate protozoa known as trypanosomes. In Central
and South America *Trypanosoma cruzi* carried by blood-sucking
triatomid bugs causes Chagas' disease, which is said to affect 7 million
people. The acute form occurs mainly in children and shows itself by
fever and various disturbances of the heart and of the cerebrospinal
system. The chronic disease is found in adolescents and young adults
and is characterized by heart trouble. The vector bugs are nocturnal in
habit and rest in the cracks and crevices of the poorer types of housing
in the daytime. Such animals as armadillos, opossums, bats, rats, cats,
and dogs serve as reservoirs of infection and this makes eradication of
the disease virtually impossible. No drug has recognizable value in its
treatment and control of the vector by the use of insecticides in houses
is the only feasible method of combating the disease.

In Africa two other species of trypanosomes, transmitted by tsetse
flies (genus *Glossina*) are responsible for sleeping sickness in humans
and in animals. *T. gambiense* is the cause of human sleeping sickness in
West Africa and appears not to occur in animals. It shows itself slowly
by generalized weakness, wasting, lethargy, fever, lymph-gland enlargement, and eventual impairment of the central nervous system. It is
often fatal. This disease is mainly carried by the riverine species of
tsetse flies, *G. palpalis* and *G. tachinoides*. A more acute form of the
disease, which often proves fatal before the nervous system is affected,
occurs in East Africa. It is caused by *T. rhodesiense* which is transmitted
by the savannah-dwelling *G. morsitans, G. swynnertoni,* and *G. pallidipes.*
This species of parasite occurs in both wild and domestic animals.

Certainly of greater importance to a greater number of people is
trypanosomiasis of domestic animals, particularly cattle. This, according
to Ansari and Junker[1], is a serious obstacle to the development of 16
million square kilometres of fertile land which could support some
125 million cattle. A minority take another point of view: if such land
were occupied by cattle controlled by the uninitiated it would lead to
overgrazing and soil erosion. The ideal compromise would appear to be
rational mixed static farming on the lines normally pursued in developed
countries. But this would mean changing a long-established nomadic or
semi-nomadic way of life for many Africans.

Trypanosomiasis, both in man and in animals, can be treated with
drugs. One drug, pentamidine, is of some value as a prophylactic against
*T. gambiense.* The insuperable problem is to locate the cases. Greater
hopes lie in the control of the vectors, which fortunately remain
susceptible to such insecticides as dieldrin, BHC and DDT. These,
particularly the first, are used to spray vegetation in areas inhabited by
the particular fly species. After the spectacular success achieved in the
1950s and 1960s in the United States in the control of the cattle screw-
worm, *Cochliomyia hominivorax*, by the mass-release of sterilized flies,
attention is now being turned to similar genetic techniques for the
control of the tsetse. This insect would seem particularly suited to this
type of control for it has a low reproductive potential and slow ability
to recover from drastic reductions in population numbers.

**Onchocerciasis**

The symptoms of onchocerciasis include itching rashes, thickening and
depigmentation of the skin, skin nodules (which contain adult worms),
and eye lesions that often lead to blindness. Possibly affecting as many
as 20 million people in Africa and Central and South America, this
disease is caused by a nematode worm *(Onchocerca volvulus)*
transmitted by various species of the genus *Simulium* (blackfly). In
Africa its distribution extends from Senegal and Angola in the west
to the Sudan, Ethiopia, and Tanzania in the east. In Latin America
it occurs in Guatemala, Mexico, Venezuela, and Colombia. The larva of
the insect vector breeds in water which has a high oxygen content and
is usually fast flowing. It normally occurs in high density, can be an
intolerable biting nuisance, and has an extensive flight range (as far as
150 km). Apart from the human misery and suffering which this disease
causes, it has led to the abandonment of large tracts of highly fertile
land and the substitution of meagre poverty for a comfortable
subsistence economy.

One of the worst endemic onchocerciasis areas of the world is the
Volta River Basin, which covers an area of 700 000 $km^2$. Of the 10
million people living in this area, one million are afflicted; 70 000 of
these are either blind or suffer severe impairment of vision. Fortunately,
a scheme is now in existence sponsored by the UNDP, the FAO, IBRD,
and WHO to provide assistance to the governments through whose
countries this enormous river flows (Dahomey, Ghana, Ivory Coast,
Mali, Niger, Togo, and Upper Volta) in controlling the disease by the
use of larvicides. The scheme is to cost in the region of $120 000 000

over a period of 20 years, the time estimated for all existing worm
infections to disappear.

Two drugs, suramin and diethylcarbamazine, can be used to cure
infections, but only under medical supervision. One of the problems is
the adverse reaction of the patient to the presence of dead worms.
Suramin kills adult worms and also has some effect on the microfilariae;
diethylcarbamazine has no effect on adult worms.

Insecticides are the only really effective means of control and some
spectacular successes have already been achieved. The best method
seems to be the application from the ground of emulsions containing
DDT to water upstream from the breeding areas. Optimum quantities
are those maintaining a concentration of 0.03 to 0.5 parts per million
(ppm) for a period of 30 minutes; they have to be related to the rate
of water flow. Such concentrations have proved non-toxic to fish.

One successful example is the control of the main African vector,
*S. damnosum*, below the Jinja Dam at the exit of the River Nile from
Lake Victoria in Uganda. Here, 12 applications at 7-day intervals at
0.036 ppm/30 minutes produced eradication for approximately 4 years
at a time and protected a population of 225 000 living in an area of
4144 km$^2$. Another example is the virtual elimination from an area of
over 10 000 km$^2$ of Kenya of the species *S. neavei,* again using DDT
as a larvicide.[8] Fortunately, no insecticide resistance has ever been
recorded in onchocerciasis vectors. This method of insecticide application is, however, potentially dangerous environmentally, and for the
long-term Volta River scheme other less persistent insecticides such
as temephos (Abate) are being considered.

### Other vector-borne diseases

Besides schistosomiasis there are a number of other helminth
diseases which depend on snails for their transmission. The lung fluke
(*Paragonimus westermani*) of the Far East — and in particular of
Japan and Korea, where the incidence may be as high as 44 per cent —
spends part of its life-cycle in snails and part in crayfish and crabs. Man
is infected when eating the raw or undercooked crustaceans. Two
species of liver fluke are important parasites of man: the cat liver fluke
(*Opisthorchis tenuicollis*), which is very common in parts of Russia,
particularly Siberia, and the Chinese or Oriental liver fluke (*Opisth-
orcis sinensis*), which occurs in Japan, China, and Indo-China and
affects up to two-thirds of the population in the worst affected areas.
Both species infect snails and fish, and man is infected when he eats

the fish. There are no satisfactory specific treatments for these fluke diseases.

A fourth helminth infection, the Guinea Worm (*Dracunculus medinensis*), is a subcutaneous infection; it is widespread in Africa and Asia and is also found in Central and South America. Up to 90 per cent of human populations may be infected in parts of India. Adult worms discharge their young, usually through leg ulcers, into water. The young then enter a small aquatic crustacean (genus *Cyclops*) and man acquires the infection through drinking contaminated water. Insecticides have been used successfully to kill *Cyclops* and so interrupt transmission.

A number of vector-borne diseases remain to be considered. They are not the great endemic diseases just described, but in normal times are sporadic and only mildly endemic. Major changes in the environment, produced by urbanization, industrial development, mass migrations in times of war, famine, and climatic catastrophes can lead to eruptions of vast proportions of some of them. Many of the arthropod-borne viruses or arboviruses transmitted by mosquitoes, ticks, sandflies (genus *Phlebotomus*), and biting midges (genus *Culicoides*) are of this nature.

Some 75 of these viruses cause disease in man. Perhaps the most notorious is the *yellow fever* of Central and South America and Tropical Africa, transmitted by culicine mosquitoes. Like most other arboviruses, it maintains itself in animal reservoirs (in this case jungle monkeys) and is transferred to village and town dwellers by forest workers, different species of mosquitoes being involved in the transmission in the different ecological situations. According to Gordon Smith,[6] 200 million people are at risk. He cites examples of recent epidemics, such as that in 1961 in Ethiopia when 30000 cases occurred with 2000 deaths. The yellow fever vaccine is most efficient and can last for 17 years and more, but to vaccinate all the people at risk would be impracticable, even though this would probably only be necessary twice in a lifetime.

*Dengue*, or breakbone fever, transmitted by similar mosquitoes is a much less serious disease but can affect thousands of people. In the famous Brisbane epidemic of 1905, 90 per cent of the city transport workers and hospital staff were out of action, and over a million people were affected in the United States in 1922. Dengue haemorrhagic fever is a more serious form, particularly among children of 3 to 6 years of age, where the mortality may be as high as 10 per cent; 8000 cases of this disease occurred in Bangkok in 1962. Both yellow fever and dengue can be controlled by measures directed against the mosquito vectors, usually in the form of insecticides. But such measures are really feasible only in towns.

Other mosquito-borne virus diseases include *chikungunya* and *O'nyon-nyong,* which are dengue-like diseases. The former is common in Africa and Asia and caused 380000 cases in a population of 1.8 million in Madras in 1965. The latter gave rise to an epidemic of some two million cases in East Africa between 1959 and 1962 affecting as much as 70 per cent of the population. The encephalitides or virus diseases that affect the brain and produce such conditions as 'sleepy sickness' are particularly dangerous diseases and have been responsible for high mortalities.

Among bacterial diseases, fly-borne typhoid and dysenteries should be mentioned. The protozoan amoebic dysentery may also be carried by domestic flies. Perhaps the best known of the vector-borne bacterial diseases is plague, transmitted to man by the bites of fleas which have acquired the bacilli from infected rats. Epidemics and pandemics of this disease have been occurring over the last 15 centuries. Some people think that its steady decline in recent years may only be a natural cyclic phenomenon and that serious outbreaks may still occur. As with trypanosomiasis, most countries maintain an expensive surveillance organization monitoring the state of the disease.

There remain two diseases the prevalence and importance of which are difficult to assess. The first is a protozoan disease carried by sandflies of the genus *Phlebotomus* and called *leishmaniasis.* The visceral form *(kala-azar),* mainly confined to parts of India, Kenya, and the Sudan, can be serious and sometimes fatal. The cutaneous form (*oriental sore*) of the Middle East is not so serious but can be disfiguring. The muco-cutaneous form of Central and South America is very dramatic in that it involves a rapid virtual eating away of the mucus membranes of nose, mouth, and throat and is often fatal. The second is a nematode disease carried by flies of the genus *Chrysops*, called *loiasis*. This is a chronic disease characterized by fugitive (Calabar) swellings caused by worm migrations. It is widely distributed in tropical West and Central Africa.

*        *        *        *

In general, then, most vector-borne diseases are tropical, rural, and most prevalent in developing countries with low standards of hygiene, sanitation, and education. It might in fact be said that these diseases and other communicable ones are in some measure responsible for the lack of development and that only with their removal or control will development proceed with any speed. The great endemic diseases are seldom dramatic in their symptoms or accompanied by obviously high mortalities. They are chronic and debilitating diseases, often overlooked, so that such morbidity and mortality data as are recorded are almost certainly gross underestimates of real prevalence. The situation is further complicated by the fact that more than one disease is often present in one individual, and may be (and often is) combined with a state of malnutrition which may itself be caused by the incapacitating effect of the diseases – a veritable vicious circle.

It is only now that the real economic effects of these diseases are beginning to be considered, because it is realized that priorities have to be made in allocating the limited resources given to public health programmes. And it is only now that the pathological changes brought about by the diseases are being investigated in detail and related to the performance of the simplest tasks, even those necessary for a mere subsistence economy. Even when all this is known, priorities will be difficult to assess. Health and well-being cannot always be measured in economic terms. What should be obvious to the countries of the developed world is that it is in their own interests to improve the health of people in the developing world, not only to reduce the chances of spread of some of their diseases (which is already happening through the promotion of tourist travel), but to produce greater equality of health and global stability.

Most of the diseases considered are related to water, and herein lies the predicament of competition between water-dependent, increased industrial and agricultural productivity, and increased prevalence of disease. Many of the man-made lakes of the tropics, by-products of energy improvement schemes, have exacerbated the prevalence of such diseases as schistosomiasis, malaria, and onchocerciasis. Irrigation schemes designed to increase agricultural output have done the same. Uncontrolled urbanization itself has contributed to an increase in disease, too, as we have seen in the case of filariasis. Thus, careful consideration needs to be given to the dilemma that the control of the disease will give rise to increased population growth, which in turn will present the need for industrial and agricultural development to feed and support an increased population with concomitant urbanization.

In only one of the vector-borne diseases, yellow fever, is vaccination at all successful, and its application is limited by the vast size of the population at risk. And few of the diseases can be controlled on a mass scale by the administration of drugs. Many could be reduced and even eradicated if education levels among the people involved were raised and simple health education instructions followed. As things are at present, however, the only feasible remedy is one of large-scale prevention by the use of insecticides or molluscicides. But these must be applied in such a way as to minimize environmental contamination.

### References

1.  ANSARI, N. and JUNKER, J. The economic aspects of parasitic diseases. *Biotech. and Bioengg. Symp.* **1**, 235 (1969).
2.  BRUCE-CHWATT, L.J. Insecticides and the control of vector-borne diseases. *Bull. Wld Hlth Org.* **44**, 419 (1971).
3.  ——, Resurgence of malaria and its control. *J. trop. Med. Hyg.* **77**, 62 (1974).
4.  FENWICK, A. The costs and a cost benefit analysis of an *S. mansoni* control programme on an irrigated sugar estate in northern Tanzania. *Bull. Wld. Hlth Org.* **47**, 573 (1972).
5.  FORSYTH, D.M. and BRADLEY, D.J. The consequences of bilharziasis. Medical and public health importance in north-west Tanzania. *Bull. Wld Hlth Org.* **34**, 715 (1966).
6.  GORDON SMITH, C.E. The role of virus diseases in tropical public health. *Trans. R. Soc. trop. Med. Hyg.* **65**, 73 (1971).
7.  LEPES, T. Present status of the global malaria eradication programme and prospects for the future. *J. trop. Med. Hyg.* **77**, 47 (1974).
8.  McMAHON, J.P. A review of the control of *Simulium* vectors of onchocerciasis. *Bull. Wld Hlth Org.* **37**, 415 (1967).
9.  ONYANGO, J.H. The present distribution of human trypanosomiasis in Africa. In *Health and disease in Africa; Proc. 1970 East African Medical Research Council Scientific Conference,* 105 (1971).
10. PAMPANA, E.J. and RUSSELL, P.F. Malaria: A world problem. *Chronicle Wld Hlth Org.* **9**, 31 (1955).
11. SOUTHGATE, B.A. and BRYAN, J.H. An investigation of the transmission potential of ultra low level *Wuchereria bancrofti* microfilariae carriers after diethylcarbamazine treatment. *Proc. 9th Int. Congr. trop. Med. Malaria* **2**, 119 (1973).
12. WHO. Mortality trends and prospects. *Chronicle Wld Hlth Org.* **28**, 529 (1974).
13. WHO. WHO Expert Committee on Filariasis. Third report. *Wld Hlth Org. tech. Rep. Ser.* 542 (1974).
14. WHO. The work of WHO, 1974. Annual report of the Director General. *Official records of the World Health Organization,* 221 (1975).
15. WRIGHT, W.H. A consideration of the economic impact of schistosomiasis. *Bull. Wld. Hlth Org.* **47**, 559 (1972).
16. ——, *et al.* Tropical health: A report on a study of needs and resources. *National Research Council Publication* **996** Washington, National Academy of Sciences (1962).

# 4 Finance for agriculture

by D.J. Ansell

There are many areas of the world where the supply of food is
inadequate. To make up this shortfall is a daunting task for world
agriculture. The rate of increase in world population, and
particularly the increase in the population of developing countries,
makes the task even greater (see Chapter 2). During the past fifteen
years the increase in agricultural output in the developing countries
has barely kept pace with population growth, despite many
technological advances. As populations continue to grow, the surplus
land capable of being used for agriculture become ever smaller, and in
some areas, notably Asia, the land/labour ratio is already low. Adequate
increases in food production can therefore be achieved only by the
intensification of farming systems and this requires capital. This
chapter analyses some of the problems of financing the development
of world agriculture.

## Capital in agriculture

Most people if asked for a definition of 'capital' would answer in terms
of the financial resources available. The economist regards as capital
certain commodities which have been manufactured by human activity
and add to productive capacity. Thus, in agriculture a farmer's capital
includes not only his implements and machinery, but also the buildings,
drainage ditches, growing crops, livestock, and stock of feed, seed,
fertilizers, and other farm inputs. We usually measure these goods in
financial terms, because money enables us to add together the value
of a number of different types of physical goods and produce an
estimate of their combined value in a single figure. Although it is a
useful measure, it is not an accurate one, particularly when rapid
inflation is taking place.

   Money is also a store of value. Thus, farmers may accumulate money
in order to purchase capital equipment in the future. Strictly speaking,
however, such reserves should not be regarded as capital: they only
become so when used to purchase productive goods such as hoes, seeds,
and fertilizers. Clearly, then, no agricultural production, however

simple the technology, can take place without the use of capital. Indeed, even hunters and fruit gatherers, in so far as they manufactured weapons and containers, used capital in the pursuit of their subsistence.

Farming systems throughout the world vary in the proportion in which capital is mixed with labour, management, and natural resources. So, in situations where land is abundant and labour cheap, one would expect to find low levels of capital investment per farm, but there are few parts of the world where this is now the case. European farmers have high costs for both labour and land; and in other parts of the developed world labour is expensive even if land is not. In the developing countries of Asia, land is generally scarce though labour is cheap, and it is only in parts of Africa and South America that land is abundant and labour cheap. If agricultural production is to continue to expand, therefore, it must do so mainly through the intensification of production rather than the extension to new land.

The form which intensification takes will vary from place to place – principally according to climate, the availability of factors of production, and market opportunities. In most of the developed countries, where both land and labour are expensive, there has been economic pressure to use techniques which produce high yields per hectare and reduce labour costs per unit of output. Thus, one observes the use of pesticides to prevent crop losses through destruction and competition, and at the same time high degrees of mechanization to reduce labour costs. In developing countries, where labour is cheap and indeed unemployed or underemployed, such development may be inappropriate. The introduction of highly mechanized agriculture into such a situation would exacerbate some of the most serious problems, even if it solved others. However, a different case can be made for the introduction of techniques which will economize in the use of land by raising crop yields. Higher-yielding varieties, fertilizers, pesticides, and, in some circumstances, irrigation are among the purchased inputs which will make the biggest contribution. The introduction of mechanization will generally raise output per hectare (tractor ploughing may result in a better seedbed, for example). Capital, then, will occupy a place of growing importance in crop and animal production. This is likely to lead to an increased demand for financial assistance from outside the sector.

Most businesses encounter problems from time to time when the flow of payments exceeds the flow of receipts. Consequently they borrow, and financial institutions exist in most countries to enable them to do so. In addition, funds for the purchase of capital may be required if businesses are to expand and if they are unable to finance such expansion from their accumulated profits.

The term 'farming' embraces a variety of different activities. Even within one country there are wide divergences in scale and in the type of farming system operated; and in a world context, generalizations might appear to be valueless. Nevertheless, many types of farm business suffer from the same type of financial problem. There is an irregular flow of receipts round the year, and frequently they are unable to generate a sufficiently large and regular flow of profits to finance expansion. For instance, a European farmer may purchase inputs for a cereal crop in the autumn of 1975 but may not actually sell the crop until the summer of 1977. Although the production cycle is essentially of one year, the time-lag between the start of an outward flow of payments and an inward flow of receipts may be twice as long. Again, many thousands of smallholder farmers in Africa are incapable of providing more than a very small margin over the subsistence requirements of their families. In order that both difficulties should be overcome, farmers need financial help.

Sometimes it may be possible and desirable to help the producer by providing him with inputs free or at a subsidized rate. This is often done, either with grants or with subsidies of various kinds. But this is an expensive solution from the point of view of the government and the tax-payer and may give rise to a misallocation of the nation's capital by promoting a wasteful use of inputs. Nevertheless, in most countries governments do assist agriculture in this way, often by enabling farmers to purchase fertilizers at concessionary prices.

Financial help can also be given by manipulating product prices. For an agricultural product which is partly imported and partly produced at home, for example, an import tariff will tend to raise the price in the domestic market. Alternatively, governments can guarantee farmers a certain price for the coming season, and either offer to buy produce at that price if farmers cannot sell it elsewhere (as is done in the EEC) or else offer to pay farmers the difference between the guaranteed price and the price which they actually receive in the market (this was the system operating in the UK before it entered the EEC). Almost all the developed countries and many developing countries use one or more of these devices for improving the financial status of their agricultural industries, but there are limitations on their ability to do so.

44

An alternative to the grant or subsidy, for some purposes, is the provision of credit. Credit is preferable to a grant or subsidy in that it is repayable and will be used only by those who need it. For the farmer, however, the interest repayments are a disadvantage and he may sometimes become involved in an ever-increasing burden of debt. Such problems will be discussed more fully below. The credit requirements of farmers can be classified according to the duration involved; usually at least three periods can be identified: short-term credit, which is normally required for less than the length of time the production cycle takes to complete; medium-term credit, probably about five years for the purchase of farm machinery; and long-term credit for the purchase of land, fixed equipment, and other long-lived assets.

Many problems exist in the creation of credit schemes for farmers which have low administrative costs and which provide farmers with a reliable source of funds at fair rates of interest. These problems are particularly severe in the poor countries. The remainder of this chapter is concerned with the problems of providing credit to farmers, and the main methods which are available, in both the developed and the developing countries.

## Developed countries

The principal problem for farmers in the developed countries is the slow rate of growth of demand for food and the more rapid rate of growth of output. The demand for food products (which constitute the majority of agricultural output in the developed countries) depends on the number of people, their level of *per caput* income, and the proportion of their income they spend on food. In North America, Japan, Western Europe, Eastern Europe, and the U.S.S.R. the rates of population growth vary between 0.6 per cent per annum and 1.1 per cent. In recent years, the rates of growth of income were impressive in the developed countries until 1974, but the rates of growth of *per caput* food consumption were low, in most cases substantially lower than 1 per cent per annum. When the likely population growth rate and and rate of growth of consumption per head are combined to form an estimate of the growth rate of demand, it is clear that in most parts of the developed world demand will increase at only about 1.5 per cent per annum. Supply, on the other hand, increased during the period 1952-4 to 1969-71 at an average of 2.6 per cent per annum for the developed countries as a whole.

The effect of this imbalance between supply and demand is that the prices of farm products tend to lag behind the prices of other products, and the incomes of farmers tend to lag behind the incomes of other sectors of the population. Thus, there arises the need for financial assistance.

In addition to the financial problems created by the slow rate of growth of farm incomes there is the fact that agricultural land has become scarce in developed countries and, as labour has moved into other sectors, the amount of capital needed per farm has increased. Leaving aside money to buy agricultural land, the level of gross investment on farms in OECD countries increased by 30-50 per cent over the last decade. Gross investment may be distinguished from net investment by the annual depreciation of existing capital assets. Of total gross investment in agriculture, purchases of machinery usually rate as the most important item; indeed, in some countries it amounts to 80 per cent of the total. Net investment is, however, much less — indeed, in many developed countries net investment is approaching zero. This does not necessarily imply an inability to raise funds or an unwillingness to invest in profitable opportunities; rather, it reflects the highly capitalized nature of farming in some developed countries. In many parts of Europe, however, where there are still small, unprofitable farms, investment is urgently needed to allow farm enlargement and modernization to take place. Inevitably in those areas where the need for finance is greatest, the availability of funds is lowest.

If one includes the purchase of land as a capital item, the finance of agricultural expansion becomes more expensive. The cost of acquiring agricultural land in most developed countries bears no relation to the profit that can be obtained by farming, but is much greater. However, it is dangerous to paint a picture of investment in agriculture with too broad a brush for there is enormous variation between countries. Thus, a recent study showed that the value of capital (including land) *per hectare* of agricultural land varied between $7500 in Japan and Belgium, $200 in Canada, and $500 in the U.S.A. and Turkey. But if one calculates the amount of capital *per farm,* a very different picture emerges, showing total investment per farm to be very high in Canada, the U.S.A., and the U.K. and very low in the countries with small farms such as Japan, Italy, and Norway.

From these figures one might conclude that the inappropriate structure of farming in many countries is contributing to the high degree of capitalization. The smaller the farm, the higher tends to be the level of investment per hectare. These large capital inputs, when compared with the poor returns experienced by many of the smaller farmers, mean that rates of return on capital are often very low, indeed frequently below 5 per cent of net farm assets. Nevertheless, most farmers inherit their land, and therefore have the use of an almost free resource; as they are reluctant to leave the industry and invest their funds elsewhere, new entrepreneurs and capital find difficulty in entering the sector. This is one reason why new capital entering agriculture has been concentrated on those enterprises where land costs are low, particularly intensive pig and poultry enterprises.

There are many sources of agricultural finance, but there may still be 'gaps' in particular countries for particular kinds of credit. The short-term, seed-time-to-harvest type of loan has traditionally been provided by the merchant or trader with whom the farmer has an intimate business relationship. The merchant knows the farmer well, can judge his creditworthiness, and can thus lend at a rate of interest which matches the risk of the loan or decline to do so. The objection often raised to this type of 'trade' credit is that the interest rates charged are high, and the farmer may find himself in a position where he is pledged to sell his crop to a particular merchant and is thus potentially exploitable. While this may be so in some cases, it is also true that most of these loans are 'unsecured' in the sense that the lender has no claim to particular, nominated assets if the farmer cannot repay. The commercial banks and other more specialized financial institutions are normally responsible for medium and long-term loans.

## Developing countries

Agriculture is said to have six main functions in the economies of developing countries:

(1) to supply food to consumers in both the agricultural and the non-agricultural sectors;

(2) to supply raw materials for industrial processing (e.g., rubber, fibres, and cotton);

(3) to transfer surplus labour to the non-agricultural sector;

(4) to supply capital funds to other sectors of the economy, either voluntarily through savings or compulsorily through taxation;

(5) to provide export earnings and to save foreign exchange by import substitution; and

(6) to act as a market for the products of industrial manufacturers.

Every developing country is different in the extent to which these functions apply. Agricultural development will in general play a crucial role and should receive a high priority in the allocation of financial resources. In the past, agriculture has often been neglected, as when governments have been tempted to divert resources to prestige industrial projects. What is more, as is implied in (4) above, agriculture has often been used as a source of finance (usually through the taxation of export crops) for the development of other sectors. This is inevitable where there are few other sources of finance available to government and where the requirements for government expenditure are endless, not only for the development of industry but also for the provision of 'social overhead capital' – i.e., roads, education, health services, and many other services. It is now, however, becoming recognized that the development of the rural sector is vital and that more of the tax revenues raised by governments should be ploughed back into it.

The emphasis in the allocation of aid from developed to developing countries is increasingly upon rural development. The total aid (about \$9 per head of population living in developing countries) from OECD countries in 1973 was \$24 150 million. Of this \$11 934 million was in the form of official aid from government to government or from multinational organizations such as the International Bank for Reconstruction and Development (IBRD), and the remainder was from private sources of various kinds. It was estimated in 1973 that about \$1500 million of the total aid budget was destined for agricultural purposes, an increase of over \$500 million compared with the previous two years. These funds are mostly destined for large agricultural development schemes, such as irrigation facilities, or are in kind – namely, tractors and other farm inputs which are consigned to governments in developing countries. Rarely, do the international aid donor and the indigenous peasant farmer come into direct contact. An increase in the absolute level of aid funds (which may be grants or soft loans) is doubtless needed so that agriculture departments in developing countries can extend their range of activities. But in many ways a more difficult problem is to devise ways in which funds which are available in developing countries can be used to assist the individual small farmers.

As in the developed countries, farmers in the developing world can be assisted financially in two main ways. First, there are government policies which might guarantee prices, restrict competing imports, subsidize the price of inputs, and make grants for various types of farm improvement. Second, there are credit schemes.

Few developing countries have the resources to subsidize their agricultural sectors by guaranteed prices. It is true that there are often restrictions on the imports of foods and that these will usually increase the price which domestic producers receive, but this is frequently matched by the high rates of taxation which governments impose on agricultural products for export. Again, inputs are subsidized, particularly fertilizers and machinery hire, but this leads to certain problems. First, in many countries there is an increasing gap between more or less static subsidized input prices, which producers have to pay, and rapidly increasing world prices, a gap which has to be filled by the government and thus the taxpayer. Second, if there is an expansion in the use of modern inputs and they are subsidized, the country is then faced with an ever-increasing burden. The better established the subsidy is, the more difficult it becomes for the government to abandon it and force producers to pay the full price. There is always the additional danger that such policies will encourage profligacy in use. In these circumstances the existence of credit facilities for farmers constitutes an important alternative source of finance for agriculture.

There is considerable risk attached to lending to small peasant farmers, particularly when they do not own their land. There are few assets against which loans can be secured, and the technical and managerial skills of the farmer are largely an unknown quantity. For this reason the most important source of loans for farmers is often other members of the family, the local trader or merchant, or professional but small-scale money-lenders. These groups usually account for the majority of short-term loans in developing countries. Interest rates are usually high, indeed sometimes they have been estimated as being as much as 100 per cent if calculated on an annual basis. For this reason such sources are criticized, but criticism is valid only if the risk element is taken into consideration and if some other institution can profitably make loans at lower rates. The latter may be difficult if we assume that accurate knowledge of individual farmers is a prerequisite in establishing their creditworthiness.

A contribution to agricultural development is more likely to be made by the provision of credit institutions which are intended permanently to raise the productivity of individual farmers. Indeed, if this is not achieved the farmer's position has actually deteriorated. If credit funds do not have the effect of raising a farmer's output, then he will be worse off by the amount of interest he has had to pay. Farmers not infrequently become enmeshed in an ever-increasing volume of debt

with little prospect of breaking free from it. The introduction of
credit schemes will therefore achieve little if they do not make
available new opportunities for the farmer; and they will be most
effective if they enable him to remove the most limiting constraint on
his business. If it is land that is scarce (as in parts of Asia), credit will be
needed for the purchase of improved seeds, fertilizers, and pesticides or
for the installation of irrigation equipment. If labour is short, credit
will be required for the purchase of machinery. Frequently, of course,
both will be required.

There are five main types of agency which can be set up to provide
additional credit channels:

(1) agricultural and development banks;

(2) farmers' cooperatives and other types of group activity;

(3) project authorities;

(4) non-institutional commercial lenders; and

(5) institutional commercial lenders.

### Agricultural and development banks

These institutions are commonly found in countries which have
undertaken a substantial amount of development planning. Their most
important characteristics, as lending agencies, are that they are highly
centralized and that in practice — although not in theory — loans are
made principally to the larger farmers. The development banks are
usually mainly or entirely state-owned. They often lack the flexibility
and local knowledge needed to make them effective lenders to
small farmers.

### Farmers' cooperatives

One way of solving the problem of the high costs involved in
lending to small farmers is to encourage them to group together and
establish a 'pyramid' of responsibility. Thus, farmers may join together
in a cooperative society which will apply to a cooperative union for a
loan. The cooperative union then approaches the development bank or
other lending institution. Although the loans are distributed among
individual farmers it is the union which is responsible to the lender and
the society which is responsible to the union. In practice, many
cooperative credit schemes have failed as a result of poor administration
or of dishonesty.

*Project authorities*

Whenever there are settlement schemes, irrigation schemes, or similar development projects there will normally be a project authority which has supervisory powers over the farming members. Because there is a degree of supervision and control it may be possible for the authority to extend credit with some confidence of obtaining repayment.

*Non-institutional commercial lenders*

If the supply of funds to existing local credit sources can be increased, it may be possible to harness their knowledge of individual farmers and at the same time increase the total flow of funds reaching small farmers. The disadvantages of this method are that there may be misuse of funds by the traders or merchants, and that in some countries it may be politically unacceptable where farmers have been exploited by local lenders in the past.

*Institutional commercial lenders*

There are cases of commercial banks lending to individual farmers, although the farmers who are able to borrow tend to be the larger ones. In most countries there is a branch network of commercial banks, and so it is not necessary to invest in new buildings and equipment. But there is still the difficulty of the high administrative costs of extending a large number of small loans to individual farmers. It may be that some association between the commercial banks and the cooperative movements would reduce costs and risks, and enable a greater mobilization of the often large resources of the banking system.

There are thus a variety of ways in which the credit supply may be augmented and improved. It should not be imagined, however, that the elimination of credit bottlenecks will solve all the problems in the development of agriculture. There may, for example, be supply problems in the provision of certain inputs, or lack of an adequate transportation system may mean that the farmer has no incentive to increase production. Certainly the small farmers need to be educated in the use of new technology and convinced of its value. Credit nevertheless constitutes one of the key elements in the process of effecting agriculture change.

*       *       *       *

There is a need to produce more food as a result of population growth and present poor nutritional standards. Most of the population increase and most of the nutritional problems will occur in the developing countries. In many of the developed countries there is a high potential for expanding food production but the domestic markets are satiated. The poor countries cannot buy from the rich as their supplies of foreign exchange are severely limited, and in any case they need to provide their own farmers with employment and markets. A central problem is to assist these farmers to raise their productivity. This will normally require extra capital, and although there are difficulties in organizing credit schemes for small farmers, this appears an indispensible prerequisite for agricultural development.

What is the future likely to hold for these farms? There is a much greater awareness of the importance of rural development and this may mean that more funds become available for their assistance. At the same time, capital will remain scarce for developing countries as a whole, and it is important that agricultural technologies are developed which reflect this fact. In many areas, a deeper understanding of the problems of small holder farmers is needed before the right kind of financial assistance can be provided. The acquisition of such knowledge is in itself expensive, but will prove a sound investment in the longer run.

### Further reading

1. ABBOTT, J.C. Agricultural credit: Institutions and performance with particular reference to the near East. *Mon. Bull. Agric. Econ. Statist.* **22** (12) (1973).
2. FAIRCHILD, H.W. Institutional farm credit systems to support agricultural development. *Mon. Bull. Agric. Econ. Statist.* **19**, 4, (1970).
3. MELLOR, J. *The economics of agricultural development.* Cornell University Press, New York (1966).
4. FAO. *The state of food and agriculture.* Rome (1974).
5. OECD. *Capital and finance in agriculture.* (2 vols.) Paris (1970).
6. TUCK, R.H. *The principles of agricultural economics,* Ch. 7. Longmans, London (1961).
7. ——, A reconsideration of the theory of agricultural credit. *J. agric. Econ.* **12** (1). 20-32 (1956).
8. UPTON, M. *Farm management in Africa,* Ch. 8. Oxford University Press (1973).
9. WORLD BANK. *Agricultural credit.* World Bank Paper. Rural Development Series. Washington, D.C. (1974).

# PART II

# STRATEGIES AND SOLUTIONS

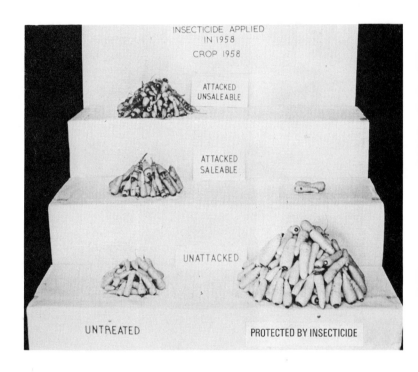

Plate 5.1. A clear example of the effect of using an insecticide to produce a market-able crop of carrots. Farmers and growers must produce not only high yields but good quality crops which the housewife will buy. (U.K. National Vegetable Research Station photograph)

# 5 The economic impact of pesticides on advanced countries

by G. Schuhmann

Pesticides are an integral part of agricultural production. Apart from their importance in helping to increase the world food supply, they safeguard the crop yields which are necessary to guarantee a minimum profitability to the farmer. The chief difference in the impact of pesticides on the developing and the more advanced countries is that in the latter pesticides are regarded as an economic input, whereas in the developing countries the question of preventing food losses takes priority.

However, we have to remember that even in the developing countries there are outstanding examples of highly profitable farm operations – and, on the other hand, that a complete collapse of the world food supply has so far been averted only by the surplus food production of the advanced countries. But it is not the aim of this chapter to compare the impact of pesticides on the developed and the developing countries: rather, it is to outline some important aspects of pest control in the more advanced countries.

More than two-thirds of all pesticides produced are used in North America, Western Europe, and Japan, and in discussing the economic impact of pesticides in these and other developed areas one has to consider the cost-benefit relationship from the viewpoint of both the farmer and the consumer. We have to distinguish between direct and indirect costs: the former are those which are borne on the farm, while others, such as investment in research, residue control, and government control of crop protection are among the indirect costs. In this chapter I have drawn principally upon examples from Germany, as that is the country of which I have particular knowledge.

## Economic aspects on the farm

### Maintaining and increasing yields

In most countries yield per unit area is the decisive criterion for profitability in agriculture: since the availability of land is a limiting factor in food production, the farmer is forced to seek the highest possible yield per hectare. However, this is not true for every advanced country. In the United States, for instance, the average yield of wheat is about 2 tonnes per hectare, which is only slightly above the world average. (The reason is that in the U.S.A. the available area is not so limited as it is, for instance, in Europe.) Conditions differ, therefore, even among the advanced countries, where yields of about 5 t/ha are quite common. In a sense, pesticides cannot produce yield increases: they can only prevent losses. On the other hand, it has to be remembered that crops including, among others, fruit, vegetables, vines, hops, oil seeds, coffee, tea, cocoa, cotton, and tobacco could not be grown profitably without pesticides. If the yield falls below the margin of profitability the farmer will be forced to give up.

In the past, crop failures, and consequently famines, were caused by outbreaks of insect pests or fungi resulting from particular weather conditions, which made it impossible for the farmer to predict his crop. The introduction of modern pesticides has radically changed this situation. Today, every farmer in the advanced countries is able to calculate within 10 per cent the size of the crop he will harvest, except, of course, after exceptionally bad weather. One example is the protection of rice in Japan, where spectacular successes have been achieved in the past 25 years. Fig. 5.1 shows that from 1949 to 1953

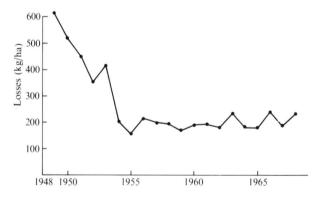

Fig. 5.1. Losses caused by insect pests, diseases, and weeds of flooded rice in Japan 1949–68. Source: ref (10).

losses from insect pests and diseases fluctuated enormously. After 1953, rice growers systematically and regularly used insecticides and fungicides, followed later by herbicides, which led to a dramatic

56

reduction of losses. As a result of this use of pesticides, and of the propagation of new varieties and better plant nutrition, Japan changed from being a rice-importing country to one with surplus rice production.[10, 19]

Another striking example of yield increase induced by crop protection is potato production in the United States from 1901 to 1974 (see Table 5.1). It is an established fact that the yield increase which started about 1941 was primarily due to the use of fertilizers, improved varieties, and irrigation. But the period since 1945 is characterized by the successful chemical control of some insects, including the potato leafhopper (*Empoasca fabae*), the late blight *(Phytophthora infestans)* and other pests. The increased yield allowed farmers to concentrate the potato crop on land best suited to potato-growing.

Table 5.1.  *Increase of yield of potatoes in the U.S.A.*

| Period | Area harvested (000 ha) | Yield (dt/ha)† |
|--------|-------------------------|----------------|
| 1901-1905 | 1262 | 62 |
| 1931-1935 | 1422 | 72 |
| 1941-1945 | 1141 | 95 |
| 1945 | 1098 | 104 |
| 1948-1952 | 662 | 161 |
| 1956-1960 | 568 | 199 |
| 1961-1965 | 551 | 224 |
| 1966-1970 | 578 | 243 |
| 1971-1974 | 540 | 276 |

Sources: refs (5) and (19).
† The decitonne (dt) has recently become the international unit of weight in agriculture (1 dt = 100 kg or 0.10 tonne).

A further example is provided by the grape yields of one of the most famous wine producers in Germany, Schloss Johannisberg (Fig. 5.2). The very accurate book-keeping at Schloss Johannisberg has made it possible to correlate yields with the factors influencing them. Many of these factors are the result of improved agricultural practices, including the use of fertilizers and new varieties. But effective control of both the downy and the powdery mildews (*Oidium tuckeri* and *Plasmopara viticola*) clearly resulted in remarkable yield responses; and the more recent developments in the control of grey mould (*Botrytis cinerea*)

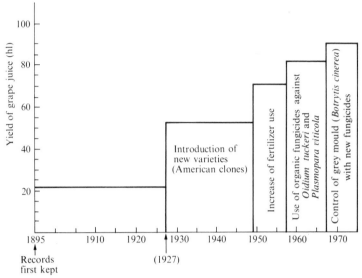

Fig. 5.2. Average yields of grape juice ('must') at Schloss Johannisberg 1895-1975. After ref. (6).

has had an additional effect on the yield. The steadily increasing yields of vines in Central Europe in general indicate that these results are not unique to one outstandingly good holding; to a great extent they are the result of using pesticides.

We can conclude that progress in crop production is the result of a number of interacting factors including the use of fertilizers, progress in plant breeding, mechanization, and irrigation. In this context crop protection provides a form of insurance: advances in breeding, irrigation, and many other agricultural practices can only be realized if they are combined with applications of pesticides.

## Improvement of quality

In all crops, and especially in fruit and vegetable growing, it is not only high yields that are required; there is also the question of quality. If we are going to consider improvement in quality, we must first define what we mean by that word. It has become fashionable, at least in Germany, to distinguish between the external and internal quality of fruit and vegetables. In theory, this distinction may be justified, but until there are some generally accepted criteria as to what 'internal quality' means, one can only use the external appearance as an indication of quality.

An experiment with two of the main apple varieties in Germany produced the average figures for 1967-75 shown in Table 5.2.

Table 5.2. *Average yields of two apple varieties in West Germany, 1967-75*

| | Cox's Orange Pippin | | Golden Delicious | |
|---|---|---|---|---|
| | Yield (dt/ha) | Marketable yield (%) | Yield (dt/ha) | Marketable yield (%) |
| 'Normal' crop protection with insecticides, acaricides, and fungicides | 162 | 85 | 256 | 80 |
| No crop protection | 95 | 35 | 167 | 25 |

Without fungicides total yield decreased by 40 per cent and marketable yield by 35 per cent.

Generally, of course, fresh fruit and vegetables cannot be sold if they show any traces of infection or insect infestation. It is well known that agricultural products deteriorate in transport and storage much more rapidly if they have previously been damaged by pest organisms: some mould fungi (e.g., certain species of *Penicillium* and *Aspergillus*) can, indeed, produce toxins which may cause disease and even death in animals and also in human beings.

*Economics of crop protection on the farm.*

An aspect of modern agriculture which is at least as important economically as preventing production losses is saving labour or making work easier. This is the reason for the constantly increasing market for herbicides, at present some 40 per cent of the pesticides market (65 per cent in highly developed countries). The interplay between manpower, modern cultivation methods, and the possibilities offered by agricultural chemicals is perhaps best demonstrated by sugar beet production in Europe. As recently as 20 to 25 years ago, the time-consuming and tiring task of hand-hoeing was still considered to be indispensable in sugar beet cultivation. Hardly anyone believed in the mid-1950s that special herbicides could ever be developed for the control of broad-leaved weeds such as fat hen (*Chenopodium album*) among the botanically closely related beet plants. But the search for such products soon met with success. Little more than ten years later manual weeding was superseded by chemical weed control which revolutionized beet-growing.

Thus, with hardly any weeds present, the purpose of hoeing became solely to single out all surplus beet seedlings; so the next step was the introduction of wide spacing and the use of monogerm seed. Instead of more than a million beet seedlings, only 70 000 to 80 000 beet plants per hectare were now sufficient to give maximum yield, provided they survived. The threat from various soil insect pests to the single-standing but still weed-free young beet plants therefore increased. So it became more urgent to control these pests, because the loss of every single young beet plant would now lead to a reduction in yield, for it could no longer be replaced by a neighbouring plant. Without this development, European sugar beet production could not have been continued on a sound economic basis.[7] Today, the hand labour required to bring the sugar beet crop to harvest, free of weeds, is less than 10 per cent of the labour required when no herbicides were used. Herbicides provide another advantage: by using them it is possible to plant sugar beet earlier than used to be possible, because there is no need for intensive mechanical weed control before sowing. Higher yields result from this earlier sowing.

It is not only sugar beet production that has benefited from the use of weedkillers. Some 85 per cent of the cereal crop in Germany is today treated with herbicides. Here, too, the prevention of yield losses is only one factor. Easier harvesting is of great importance to the farmer: in order to use combine harvesters the grain in the ear must be dry enough for it to be threshed, but heavy weed infestation keeps the cereal crop moist; in addition, a heavy weed population hampers the actual harvesting.

There have been similar developments in maize-growing. The successful introduction of this crop into European agriculture was to a great extent made possible by the use of herbicides of the simazin type, which enabled maize to be grown without mechanical weed control. It was this one factor which was responsible for the huge increases in maize production in West Germany between 1963 and 1973. In 1963 13 000 hectares were grown, yielding 48 000 tonnes; ten years later the figures were 106 000 hectares and 573 000 tonnes.

The principle of using chemicals to save labour is, of course, followed by many crops besides sugar beet and cereals: rice, cotton, potatoes, fruit, vegetables, and ornamentals are just a few that come to mind. It is estimated that, according to the degree of weed infestation, about 5 to 25 hours per hectare are needed to reduce weed competition in the U.S.A., compared with 32 to 120 hours per hectare without herbicides.[4]

There is no doubt that the use of pesticides, together with the use of fertilizers, mechanization, and plant breeding, has contributed significantly to the reduction in Western Europe of the number of people engaged in agriculture by two-thirds in the last 25 years — a reduction which has been made possible only by concentrating on a few profitable crops and by using every technical advance. This has meant, of course, that farmers have needed to be well informed.

## Comparisons of costs and benefits of using pesticides

The chief aim in using pesticides is to prevent crop losses which result from pests, diseases, and weed competition on the farm. So far as the farmer is concerned his input of pesticides must be justified by the increased value of his crop. In order to achieve this, the farmer must seek the most promising pesticide at the lowest price. To succeed, pesticides commonly have to be applied before the infection occurs or before it is evident; so early diagnosis and forecasting are vitally important.

In the developed countries it is common for farmers to receive, from government services, regular reports indicating how, when, and to what degree their crops are endangered by certain pests, and the action to be taken against them. Such a service requires, of course, a highly specialized staff.

It has to be remembered that crop protection involves not only the cost of pesticides: application costs have to be added. In Germany, based on various records, mostly unpublished, they range between 30 per cent and 90 per cent of the material costs, averaging about 70 per cent of the total cost of a chemical control operation. This leaves the farmer only about 20 per cent of the total with which to choose the pesticide which will give the best results at the lowest cost. The application costs include the speed of working, the use and depreciation of expensive equipment, and the amount of labour employed.

In advanced countries, therefore, especially in areas with very small agricultural holdings, cooperative approaches have proved to be efficient. For example, collective treatment of vineyards by a helicopter was not only efficient in itself but it reduced the degree of infection (from neighbours who might have taken no action). There was also a saving of 50 per cent on pesticides compared with conventional ground equipment.

There are, of course, various possible side-effects which can result from the use of pesticides. One is the possibility of phytotoxicity, which can occur as a result of drift or improper use of herbicides. Another factor is that of rotation in agricultural crops. So far as insecticides are concerned, it is sometimes unavoidable that some harmless or beneficial species are killed as well as the pest species which it is intended to kill. A more common event is that in reducing a pest species a beneficial predator is also reduced, so allowing a different species to become common and even a pest, such as red spider-mite in apples. Other such complications can occur.

The development of resistance to particular insecticides is well known and recently resistance of certain fungi to fungicides and thus promotion of non-target fungus diseases has been encouraged.

Some side-effects are beneficial. Thus, weed control is aimed at eradicating competition with the crop but it may also eliminate wild hosts of plant viruses and plant nematodes. Herbicides also reduce the need for tillage and so help to conserve soil moisture in dry areas like the Corn Belt and the Great Plains of the U.S.A. The cash value of such effects cannot be quantified.

One thing which is quite incalculable in the context of a cost-benefit analysis of pesticide use is the amount by which the price of food to the consumer could be reduced by higher crop yields resulting from an increase in the use of pesticides. The farmer has no choice but to keep his expenditure as low as possible and to increase his earnings as much as he can. In most of the developed countries the amount of available arable land is limited, and so the only way of producing more food is to increase yields. For this a minimum input of modern technology is essential, and is more or less constant for any particular type of farming operation. The decisive factor is the reduction of net income resulting from crop losses (Fig. 5.3).

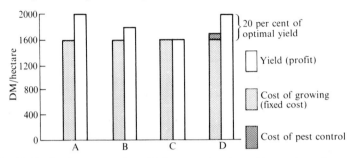

Fig. 5.3. Production costs and profits in comparison to pest losses. A, no losses and no pesticides; B, 10 per cent losses (= DM 200), no pesticides; C, 20 per cent losses (= DM 400) no pesticides; D, no losses, with pesticides (= DM 100). Adapted from ref. (14).

Therefore, on intensively managed agricultural holdings, with a high level of investment, the higher yield is linked with a higher investment in crop protection. This applies to all crops which need a high input, whether in terms of money, manpower, or both. If losses can be avoided, the profitability of other inputs, like fertilizers, can be improved.[9] Indeed, as with fertilizers, the law of minimum return applies equally to pesticides. This means that the input can increase the output only to a certain optimal level. Fig. 5.4 shows that an optimal economic result will be obtained only with a maximum distance between expenditure

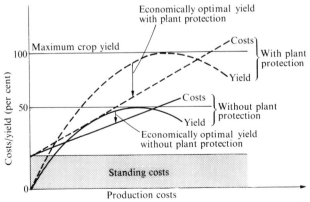

Fig. 5.4. Relationship of expenditure to returns. Source: ref. (19).

63

and return. Generally, the farmer can expect to have a return of 1 : 3 to 1 : 9 on his investment in crop protection. In special cases this ratio can increase to 1 : 100.[3]

*Examples*

It is useful to look at some concrete examples of the effect of pesticides on profitability. In the U.S.A., for instance, it has been estimated that banning the use of phenoxy herbicides without substituting another effective weed control measure would reduce average yields of wheat and small grains by 30 per cent on weed-infested land. On this assumption, for wheat farmers the average loss in income has been estimated at nearly $ 30 per hectare,[1] whereas the cost of using herbicides is only about one-sixth of that amount.

When relatively costly pesticides are used, the margin between expenses and profit and from extra yield is frequently narrow. For this reason, the chemical control of nematodes is usually not so much a technical problem as an economic one, and in many crops it is not economic to use nematicides. But in cotton fields, where losses due to nematodes are 10 per cent more, control practices usually result in yield increases sufficient to justify the cost,[22] and there are other crops of high value which will stand the cost of soil treatment. Thus, in treating strawberries against certain pests and diseases, yields of more than five and ten times the cost for treatment have been achieved. Comparable degrees of profitability can also be obtained in fumigating greenhouses. Nevertheless, it has to be remembered that the economic benefit depends very much on the level of infestation.

In most advanced countries, seed dressing of cereals is a routine crop-protection practice. The costs involved represent generally 0.25 to 1.00 per cent of the gross value of the yield. Seed dressing, therefore, is always profitable, because yield increases of 1 to 3 per cent are absolutely certain. In cases of heavy fungal infections of bunt and smut *(Tilletia, Ustilago* sp.*), Fusarium* and *Helminthosporium* diseases can cause yield losses of 50 per cent or more.

Again, with cereals, some estimates conclude that 50 to 90 per cent of the cereal-growing area needs herbicidal treatment. For example, over 6 years the application of soil herbicides gave on average an additional yield of 4.19 dt/ha. The costs of crop protection are about

DM 60/ha, which is equivalent to 1.5 dt of cereals; this means that the investment yielded a return of 280 per cent.[17]

Trials during 5 years in Western Germany assessing sugar beet losses due to aphids and the virus yellows disease which they carry have shown average losses of 27 per cent sugar, which could be prevented with insecticides. This means that for similar infested areas in Germany three applications of insecticides give a yield increase of 127 dt/hectare and a cost-benefit ratio of 1 : 8.[21]

Headley has estimated the productivity of pesticides in the U.S.A.; he found that all types of pesticides generally returned more to farmers than they cost.[8] On average, one dollar for pesticides generated four dollars of additional sales.

It is seen, therefore, that pesticides can be considered very effective tools for improving agricultural yields, and, consequently, for increasing and securing the farmer's income. What is more, prices of pesticides have remained relatively stable: as Fig. 5.5 shows, the price index for pesticides (1962-3 = 100) in West Germany reached only 105 in 1974-5 compared with an index of 153 for all means of

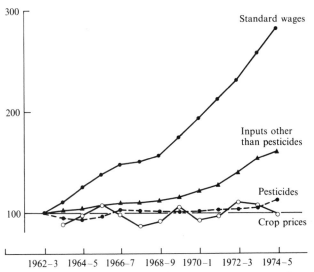

Fig. 5.5. Index numbers of prices received by farmers for crops and of standard wages, inputs without pesticides, and prices for pesticides in West Germany: 1962-3 = 100. From ref. (20).

agricultural production. Relationships of this kind apply to many countries.

### Effects on the consumer and society

*Benefit*

The quality of food offered today to the average consumer is undoubtedly higher than ever before. It has been pointed out above that plant protection minimizes the grower's risk, and this has direct consequences for food policies. The fact that yields have become calculable to a certain extent enables governments to predict the food exports and imports which their countries will need; it also safeguards food supplies for the consumer. Most consumers do not realize that many tropical and sub-tropical fruits, such as citrus and bananas, are continuously available only because they are protected by pesticides before and after harvest, in transport and in storage.

Thus, for Spain, the export of citrus fruits is important to the national economy because it is a major source of foreign exchange. This export trade may, however, be greatly affected by the quarantine measures imposed by importing countries to prevent the introduction and spread of the Mediterranean fruit fly (*Ceratitis capitata*), the most important pest of oranges in Spain. In 1964, for example, West Germany rejected 711 tonnes of citrus fruit, which was found to be infested with Mediterranean fruit fly. The Spanish Government then made it obligatory for fruit fly control measures to be carried out in the citrus plantations, with the result that the amount of rejected fruit had dropped to 14 tonnes by 1968.[12]

As is discussed in Chapter 14, pesticides play an important role in the preservation of food, by controlling insects and other animal pests in stored products. Treatment can be carried out on infested material without disturbing it, in such places as the silo or warehouse, or in transport. With the aid of pesticides it is possible to store staple food, including rice and other cereals, for years without any risk of loss due to insect pests. The expense is minimal compared with the benefit gained. Modern storage, as it is presently practised in many countries, not only contributes to the prevention of losses; it is also a pre-requisite for the continuous supply of food, since surpluses can be

stored without risk for several years. This, together with steadily
increasing yields, leads to stable prices for agricultural products.

It is estimated that in the absence of crop protection measures, and
of treatments for the protection of stored products, prices would be
50 to 100 per cent higher for foods from such crops as fruit, vegetables
and non-staple commodities, and 25 to 50 per cent higher for staple
foods. With a constant national income, this would have its effect on
purchasing power, and so on industry and the economy in general.
Stability in food prices is part of the basis of every sound national
economy, and in the advanced countries plant protection has without
doubt made a substantial contribution in this field. As we have seen,
Fig. 5.5 compares movements in wages, crop prices, prices of pesticides,
and the cost of other agricultural inputs between 1962-3 and 1974-5. It
is obvious that pesticides helped to keep crop prices stable.

Furthermore, it must be remembered that in advanced countries
industrialization and agricultural progress are connected and interact
with each other. Thus, agriculture plays its part even in industrial
society; it is an essential partner in, for example, the agricultural
machinery industry, the fertilizer industry, and the pesticides industry.
Taking the world-wide turnover of pesticides as about 15-17 thousand
million DM, it is obvious that this also has some bearing on labour
supply, taxes, and trade exchange. Cramer has pointed out that the
classical distinction between agricultural and 'industrial' countries no
longer holds, since the highly industrialized nations are today also those
with the highest agricultural output, and so contribute considerably to
the world food supply.[2]

## Costs

Any analysis of the social relevance of pesticides would be incomplete without discussing, besides the benefits, the costs and risks
involved. Possible hazards for human and animal health have to be
considered as well as possible unwanted side effects on other non-target
organisms, on the environment, and on the ecosystem as a whole, which
is the basis of every form of life on this planet. As other authors in this
book may well point out, the example of DDT, for instance, indicates
that it is quite possible, after extended use, for traces of persistent
insecticide to spread world-wide, being transported in the atmosphere

and the oceans; and that a persistent compound can be accumulated in certain parts of the food chain. Other pesticides which have a high acute toxicity can endanger human life if they are improperly used.

It is impossible to measure unwanted side-effects in terms of money, and it is necessary to do everything possible to prevent their occurring in order to diminish the risk to man and the environment. Most countries, therefore, have laws and regulations to control trade in and the use of pesticides. Usually a pesticide has to be officially tested and to be approved or registered before it can be marketed (see Chapter 19). These regulations require that pesticides shall be applied in a way which guarantees that officially tolerated residues are not exceeded, so that the consumer is protected from every calculable risk. Residue tolerances are determined by national and international bodies, both for human food and for animal feedstuffs. As these questions are discussed in Chapter 19, I wish only to emphasize here that legislation in all advanced countries has reached a standard which, given the proper use of pesticides, ensures that the consumer of agricultural products suffers no risk to his health or wellbeing. To obtain registration of a new product the manufacturer of a pesticide has to produce to the authorities extensive documentation indicating its mode of action, toxicity, residue data, and much other information. Requirements for registration are constantly being increased, and this, of course, increases the cost of research and development. This means that in the long run pesticides, at least during their patent lifetime, become more expensive to the farmer, and thus, through higher food prices, to the consumer.

Society as a whole has to pay for the cost of government control bodies and research facilities. To this has to be added the cost of running the necessary agricultural extension services. In West Germany it is estimated that the taxpayer has to find the equivalent of about 5 per cent of the cost of using pesticides to finance these services. Doubtless this applies to other countries as well.

## The future

The general aim of agriculture is the production of high-quality food, animal feedstuffs, and other crops at minimum cost. For this purpose virgin areas have been cultivated, new high-yielding crop varieties bred, plant nutrition constantly modified, and many technical advances used to improve growth conditions and harvesting operations. In this system, the role of crop protection is to minimize yield losses and also to save manpower and mechanical operations, and so help to bring farm income up to the level of industrial wages.

The control of insects and the protection of crops is essential if we are to produce enough human food and animal feed of high quality. Insects must also be controlled if we are to enjoy freedom from the many diseases which they carry to man and livestock (see Chapters 3 and 9). There is no doubt that modern pesticides are highly efficient tools which can be used to reduce most of the known plant pests and diseases to a tolerable level. Nevertheless, some scientists and members of the public are impressed by the suggestions of harm done to the environment and wish to replace pesticides by other control methods. Most scientists and economists agree, however, that for the foreseeable future pesticides will be necessary for the protection of man's food and fibres. Pesticides have become an irreplaceable and integral part of the production process that has evolved from increased specialization and more intensive farming. Banning their use would have an immediate impact on the farmer, on the consumer, and on society as a whole.

What we need more than ever (especially in regard to legislation and advisory bodies) is a balanced view of the risk-benefit relationship of pesticides. Research and development within the chemical industry should not be discouraged by the increasing burden of costs, because progress can be achieved only by encouraging, not restricting, research. It is completely unscientific to oppose 'pesticides' in general: there are so many different compounds, different in their mode of action, in their toxicology, and in their effect on the environment, that generalization means oversimplification. We now have the means of discovering the behaviour of new products and of minimizing the risk of their application to a perfectly acceptable level. In the past 30 years we have made great progress in the development of new, highly active, and relatively efficient pesticides which are reliably safe, easy to use, and cheap. There is no reason to doubt that we shall have further successes in this field in the future.

$$* \qquad * \qquad *$$

To summarize, pesticides are efficient tools for preventing crop losses due to plant pests and diseases. They enable farmers to produce regularly, good yields of high-quality crops. Furthermore, they contribute to the rationalization and intensification of farming and to more effective harvesting. As aids to production they are relatively cheap. Modern farming is, therefore, in many respects dependent on the proper use of pesticides. Many crops could no longer be profitably grown without the help of these chemicals; in other crops they safeguard high profitability. Generally, pesticides contribute to a stable food price level and are an important factor in keeping food supplies in balance. In a modern, industrialized society they are a connecting link between industry and agriculture, and are an integral part not only of modern agriculture but of society as a whole.

*Acknowledgement.* I should like to thank Dr. H. H. Cramer of Leverkusen for his close cooperation in preparing this chapter.

### References

1  ANDRILENAS, P.A. Evaluating the economic consequences of banning or restricting the use of pesticides in crop production. In *Economic research on pesticides for policy decision making. Proc. Symp. U.S. Dept Agric. Econ. Res. Service,* Washington D.C., (April 27–29, 1970).
2  CRAMER, H.H. Pflanzenschutz und Welternte. *Pfl Schutz-Nachr. Bayer, 20,* 1-521 (1967).
3.  CRAMER, H.H. Zur Wirtschaftlichen Bedeutung des Pflanzenschutzes. *In* Wegler, R. *Chemie der Pflanzenschutz-und Schädlingsbekämpfungmittel* Vol.1. Berlin and Hamburg (1970).
4  ENNIS, W.B. Restricting the use of herbicides − what are the alternatives? In Economic research on pesticides for policy decision making. *Proc. Symp. U.S. Dept. Agric. Econ. Res. Service,* Washington D.C., (April 27–29, 1970).
5  *FAO Production Yearbook* 1974, 28 *1,* 2 FAO, Rome (1975).
6  GOELDNER, H. Uber die Entwicklung des Rebschutzes in Westeuropa. *Pfl Schutz-Nachr. Bayer, 22,* 81-7 (1969).
7  HANF, M. *Von der Mechanik zur Chemie − 35 Jahre Pflanzenschutzentwicklung.* Mitt. Biol. Bundesanst. Land-und Forstwirtsch. Berlin-Dahlem, **146,** 9-37, (1972).
8.  HEADLEY, J.C. Economics of agricultural pest control. *Ann. Rev. Entomol.* **17,** 273-86, (1972).
9  HEITEFUSS, R. *Pflanzenschutz, Grundlagen der praktischen Phytomedizin.* Georg Thieme, Stuttgart (1975).
10  JUNG, H.F. and SCHEINPFLUG, H. Reisanbau und seine Pflanzenschutzprobleme in Japan. *Pfl Schutz-Nachr. Bayer, 23,* 243-71, (1970).
11  KOLBE, W. *Pfl Schutz-Nachr. Bayer,* **29,** (in the press) (1976).
12  KOPPELBERG, B. and CRAMER, H.H. Die Bekämpfung der Mittel eerfruchtfliege *Ceratituis capitata* Wie. in Spanien. *Pfl Schutz-Nachr. Bayer,* **22,** (1), 164-74 (1969).

13 ORDISH, G. Some notes on the short-term and long-term economics of pest control. *PANS* sect. A, 14, 343-55 (1968).
14 —, and DUFOUR, D. Economic bases for protection against plant diseases. *Ann. Rev. Phytopathol.* 7, 31-50 (1969).
15 PETERS, D.C. The value of soil insect control in Iowa corn. *J. econ. Ent.* 68, 483-6 (1975).
16 PRICE JONES, D. The energy relations of pesticides. *Span,* 18, 20-2 (1975).
17 RESCHKE, M. Gedanken zur Unkrautbekämpfung im intensiven Weizenanbau, *Gesurde Pfl* 27, 194, 196-7 (1975).
18 RIETH, G. and SCHULTE, G. Mehrjährige Untersuchungen über die Wirtschaftlichkeit von Pflanzenschutzma nahmen in zwei landwirtschaftlichen Betrieben Nordwestdeutschlands. *Z. Pflanzenkrankh. Pflanzensch,* 77, 309-27 (1970).
19 SALZER, W. Economic importance of chemical crop protection in relation to its ecological impact. *Environ. Qual. Saf.* 2, 271-80 (1973).
20 *Statistisches Jahrbuch uber Ernahrung, Landwirtschaft und Forsten* 1975. Bundesministerium für Ernährung, Landwirtschaft und Forsten, Hamburg und Berlin (1975).
21 STEUDEL, W. Neuere Erfahrungen zur Frage der Ertragsverluste durch die viröse Vergilbung der Zuckerrübe. *I.I.R.B.* 6, (2), 60-6 (1973).
22 WEBSTER, J.M. *Economic nematology.* Academic Press, London and New York (1972).

Plate 6.1. A heavy infestation of the weed wild oat in a crop of wheat. In recent years this weed has been a major problem in temperate cereal production, but it can be controlled with certain new herbicides. (ICI photograph)

# 6 Some problems of temperate cereal production

By K.S. George and J.E. King

Cereals are a major constituent of the diets and economies of all peoples of the world, mainly because they can be stored without much deterioration even for years. They consequently have a place in all agricultural systems, occupying more than 700 million hectares or approximately 40 per cent of the area of the earth under arable cultivation or permanent crops. Of this 40 per cent under cereals, approximately one-third is in the temperate region, devoted largely to the growing of wheat, barley, oats, rye, and maize. Maize is grown principally in the warmer, more southerly areas of North America and Asia but in recent years the breeding of cold-tolerant varieties has led to increased production of maize for silage or grain as far north as southern England, France, Holland, Germany, and Eastern Europe, including the U.S.S.R. However, the total area under maize in these countries still represents less than 20 per cent of the world crop.

The production of temperate cereals, estimated from statistics for 1973, was 944 million tonnes of grain, of which approximately 40 per cent was wheat, 33 per cent was maize, 14 per cent barley, 6 per cent oats, and 3 per cent rye. The relative proportions of total grain attributed to individual crops varied between areas. In North America maize was the heaviest crop with more than twice the production of wheat, which was the second most important cereal crop. In Eastern Europe and the U.S.S.R. wheat production was twice that of barley, four times that of maize, and more than six times that of rye and oats. In the United Kingdom, barley production was greatest, followed by wheat and then oats.

The grain produced is used for human food, animal feed, and seed. Wheats of good quality are milled to produce flour for bread or cakes and biscuits. The wheat germ, bran, and other milling products taken from the flour, together with poor-quality grain, are constituents of processed animal feeds. Barley is also used to produce compound animal feeds and, if of adequate quality, to produce malt for brewing or distilling. Rye similarly may be used for distilling and compounding but in Scandinavia and Eastern Europe much rye is made into bread flour. Oats may be made into porridge or compounded in animal feeds. Maize also provides feed for livestock but is, in addition, an important raw material for the production of sugars and starch essential to the food-processing industry. All types of grain must of course be resown annually to maintain production, and each year up to 5 per cent of the production is required for this purpose.

Cereals can be grown in many different soil types and climates and may be sown in either winter or spring according to the system of farm management, but the variety used must be appropriate to the time of sowing. Winter cereals require a period of exposure to cold winter conditions to initiate the physiological change which results in reproductive growth and the production of an ear. Consequently, they must be sown early enough to ensure such a period of vernalization. Spring varieties do not require a vernalization period and may be sown at the end of the winter but they will usually fall short of the yield of a winter variety by 15 to 20 per cent.

## Plant breeding for disease resistance

At the turn of the century, cereal crops in the U.K. were yielding in the region of 2.15 to 2.75 tonnes per hectare. In the early 1970s, average yields were in the region of 3.85 to 4.35 t/ha. This increase is attributable in large measure to the breeding of varieties which, by virtue of desirable physiological and structural characteristics, are able to utilize the greatly-increased quantities of fertilizers applied, producing heavier yields while retaining the ability to stand without lodging, a quality which both enhances yield and facilitates ease of harvesting.

Concomitantly with this objective of improved yields, the breeder must also ensure that new varieties are resistant to the many diseases which can attack cereal crops. Genetic variables which convey some degree of resistance to the diseases that are likely to be encountered in the field are consequently present in varieties at the time of their release into commerce. Where, however, the resistance to any particular disease is conveyed by a single genetic factor or major gene, it is often quickly nullified by the rapid appearance and spread of new strains of the pathogen which are adapted to overcome the effect of the single gene and to infect the hitherto resistant variety. In the past 25 years, at least eight varieties of wheat in the U.K. have been lost in this manner to new strains of the fungus which causes yellow rust *(Puccinia striiformis)*; similar failures have occurred in barley varieties released with major gene resistance to mildew *(Erysiphe graminis)*.

Breeders are therefore turning to another form of resistance, so-called field resistance, which is probably determined by many genes. Field-resistant plants are not completely immune to disease but become infected only to a restricted degree which is insufficient to affect significantly the final yield. Even so, it is sometimes necessary to release a variety which is relatively susceptible to disease but which is capable of outyielding existing varieties to an extent which precludes its being withheld from commerce. Such varieties, as with field-resistant ones, will be affected by disease to some degree which will dependend upon the environment and may become economically damaging. In such cases, as with the new high-yielding semi-dwarf wheats which are susceptible to *Septoria (S. nodonum and S. tritici),* disease-control using fungicides can be a necessary economic measure reinforcing the efforts of the breeder in improving and maintaining yields. However, it is established that some diseases produce genetic variations which are insensitive to certain fungicides and can spread and grow in their presence. Such variations may be selectively favoured by particular spraying regimes so that the breeder has constantly to introduce variations to the genetic make-up of the seed, and not rely entirely in the long term on chemical control (see Chapter 16).

### The pests and diseases of cereals

Cereal growers are faced with a variety of pests which attack plants above and below ground. A growth stage at which the plant is particularly vulnerable is between germination of the seed and the end of tillering. As soon as the seed is sown, it may be attacked by wire-worms *(Agriotes* spp.) or slugs *(Agriolimax reticulatus),* which leave an empty husk. Pheasants and rooks work their way along the rows leaving empty spaces, while, on the margins of fields, rats and mice may tunnel and destroy a small amount of seed.

When germination has taken place this damage continues, but other pests also appear, notably leatherjackets (*Tipula paludosa* and the larvae of various other crane flies), wheat-bulb fly *(Leptohylemyia coarctata)* and, mainly in oats, frit fly *(Oscinella frit)*.

The eggs of crane flies are laid in grassy, weedy fields in the autumn and the larvae hatch and feed, first on the grain and then on the developing shoot, frequently biting this off just below ground level. The larvae complete their development within 6–9 months of the eggs being laid and at maturity may be 2.5 cm long. To achieve this growth rate each eats many seedling cereals and an infestation may completely destroy large areas of the crop.

Whear-bulb fly is a very common pest, mainly on winter wheat, and like frit fly, it is a stem borer; that is, the larvae hatch and tunnel into the developing shoots. Wheat-bulb fly eggs are laid on bare soil during the late summer and may be numbered in millions per hectare. Larval attacks may completely destroy whole fields of wheat plants or, more usually, thin the crop to such an extent that the potential yield from the area is unlikely to be realized.

As the plants grow, other symptoms may occur which indicate the widespread presence of eelworms (nematodes) attacking the roots or one of many minor pests which are capable of damaging plants, usually on a restricted scale.

During the stages of stem elongation and production of the ear, aphids may colonize and breed on various parts of the plant. From early summer onwards, stems and leaves may be attacked by up to several hundred aphids per stem (mainly *Metopolophium dirhodum*); these insert stylets into the plant tissue and remove the sap, causing considerable loss in grain yield. They also act as vectors of barley yellow-dwarf virus, which can multiply in the plant and cause losses as great as or greater than those caused by the aphids. When the ears of plants emerge, another aphid species *(Sitobion avenae)* colonizes the ears and, if present in numbers of five or more per head once the grains start to fill (that is, after flowering), can cause yield losses of up to 15 per cent.

In addition to infestation by pests, temperate cereals are subject to diseases caused by many micro-organisms. These are mostly fungi, although some diseases such as halo blight of oats *(Pseudomonas coronafaciens)* are caused by bacterial pathogens and others, such as barley yellow-dwarf, affecting all temperate cereals, are caused by viruses. All the organs of the plant are subject to attack by one or more pathogens, the most common of which may cause national losses as high as 10 to 15 per cent of the yield, and can occasionally devastate individual fields.

Roots can be attacked by the take-all fungus *(Gaeumannomyces [Ophiobolus] graminis)*. Absorption of water and nutrients from the soil is then impaired and grain weight reduced. In early severe infections, 'whiteheads', which are prematurely ripened, bleached, empty ears, may result.

The stems of plants may be attacked by various foot rots, including eyespot *(Pseudocercosporella herpotrichoides)*, which impede translocation of nutrients thereby reducing yield and often weakening the stems so that they lodge (fall over).

The leaves and leaf sheaths are open to attack by mildew, yellow rust, brown rust *(Puccinia hordei* or *P. recondita)*, black rust *(P. graminis)*, and leaf spots *(Septoria* spp, *Selenophoma donacis, Rhynchosporium secalis)* which can kill much of the photosynthetic area of the plant. This results in poorly filled ears or, in the case of early attacks, retards the development of roots, tillers (extra shoots), and grain primordia in the developing ears. Most of these foliar diseases can also attack the ears, reducing yields by impeding translocation of nutrients to the grain or, in awned cereals which produce photosynthates in the ears for storage in the grain, by reducing the photosynthetic area of the plant.

Other diseases such as bunt *(Tilletia caries)*, loose smut *(Ustilago nuda)*, and covered smut *(U. hordei)* contaminate or infect ripe seed but infection is not manifest until the following season. The infected seed germinates to produce a systemically infected plant in which the fungus grows within the host, keeping pace with the ear primordium until ear emergence, when it produces its spores in place of the grain. This causes total loss of yield in that tiller. Ergot *(Claviceps purpurea)* similarly replaces grain with fungal tissue called sclerotia, although infection occurs after ear-emergence and symptoms appear later in the same season. Only a few grains are affected in each ear and the effect on yield is small; the main disadvantage of infection is the production of alkaloids by the sclerotia. These are poisonous to animals and man, and excessively contaminated grain is useless for stock feed, for human consumption, or for seed.

The summary given above will serve to illustrate the hazards to which the crop is exposed from the moment the grain is sown until the ripe grains are harvested. If any of the pests mentioned occur, as they frequently do, in optimal conditions, the crop may be completely destroyed. Usually not one but several pests and diseases occur on the same crop, but usually at a low level. Under these circumstances, catastrophic destruction does not occur but small yield losses undoubtedly take place. When pests and diseases occur together their combined effects may be more or less damaging than the sum of their individual effects, but it is impossible, with present knowledge, to apportion yield losses to any particular organism.

*Some problems of temperate cereal production*

### Pest and disease control

Very few farmers expect to sow a field of cereals and leave it until harvest. Even if symptoms of pest attack are not immediately obvious, pest damage is sufficiently well known for precautions to be taken to ensure the largest possible yield. Cultural methods of pest control have always been a feature of farming, having in the past been the only methods the grower could adopt to prevent crop losses. Whereas there is little he can do to prevent pigeons or rabbits from destroying part of his crop, he can make sure that potential insect pests are given the least favourable circumstances in which to operate. Where leys are to be ploughed in and the land used for cereal growing, this is done as long as possible before sowing so that indigenous grass pests are destroyed. Land which is bare fallowed may not be suitable for cereal growing, but if this is inevitable then winter wheat seed is sown early and shallowly to lessen damage from wheat-bulb fly. If cyst eelworms *(Heterodera avenae)* are a problem in the field, resistant varieties of cereals may be used which minimize the risk to grain yield.

Even with all these precautions, the farmer is only too well aware of catastrophes which may occur and his final resort is the chemical control of pests. Seed dressing, usually applied by the seed merchants, is regarded as a good insurance for cereal seed and its widespread precautionary use controls seedling pests such as wheat-bulb fly. Pests which occur in the aerial parts of the plants, and are usually sporadic in appearance although often locally severe, are controlled by a variety of sprays and occasionally baits.

Similar cultural practices or chemical treatments are used to control the cereal diseases, the precise methods being determined by the life-histories of the various pathogens, their mechanisms of dispersal, and their means of surviving between the end of one season and the beginning of the next. Pathogens which have a soil-borne phase in the life-cycle, during which they live on debris from previous crops, can be controlled by careful attention to rotation. In the case of take-all it is inadvisable to grow the highly susceptible wheat crop longer than one year after a single year's break from cereals or two to three years after a break of several seasons. Eyespot, sharp eyespot *(Rhizoctonia solani),* *Septoria,* and *Rhynchosporium* leaf blotches, all of which can survive on debris, are favoured by inadequate stubble hygiene and may possibly increase in importance where 'no-plough' techniques are employed.

Diseases which have a seed-borne phase, including *Septoria,* the *Fusarium* footrots, seedling blight and ear blight diseases, bunt, and also covered smut, and leaf stripe of barley *(Pyrenophora graminea)* are controlled by organo-mercury seed dressings applied before the seed is sown. Loose smut can be controlled by hot water treatment or seed dressing.

The control of diseases like mildew and the cereal rusts was once entirely dependent upon the growing of resistant varieties, but in recent years fungicides active against them have been developed. It is possible to control these diseases by sprays applied to crops during the growing season. In the case of barley mildew, fungicidal sprays or seed dressing are effective measures. Control by sprays may also be extended to the air-borne or rain-splash dispersed phases of the leaf-blotch diseases, *Rhynchosporium* and *Septoria,* or the stem-base disease eyespot.

It must, however, be stressed that the farmer often needs advice on the cost-effectiveness of sprays, for the relatively low value of the crop does not justify the use of chemicals as a routine precaution unless the degree of infestation is such that positive benefits can be demonstrated. Whether or not to spray and when to spray are questions which require careful judgement, by farmers and by the advisory and extension services, of the likely effects of a spray on yield, taking account of the climatic conditions and the likelihood of their allowing a particular pest or disease to develop to an extent where the damage it might cause would be economically worth controlling.

The time and expense involved in answering these questions and the need for farmers to keep a constant watch on their crops are dis-advantages, as is the environmental hazard which must occur whenever toxic chemicals are released into the biosphere. However, costs of materials and application discourage indiscriminate use of chemical sprays, and modern practices, correctly followed, tend to minimize the risks of pollution of soil, waterways, and the habitat in general, and to have little potential risk for other wildlife.

The use of fungicides and pesticides has diminished the depredations of soil and seed-borne pests and diseases in the past and it seems certain that chemical sprays, which often are the only effective means of preventing large yield losses, will enhance the control of foliar pests and pathogens in the future.

## The effects of weeds on cereals

Routine spraying of cereal crops against weeds has been practised on most of the cereal crops in temperate climates for over twenty years.

Investigations of their effects have tended to be concerned with weed ecology rather than with the yield benefits of treatment. The use of herbicides is, however, part of the overall programme of cereal production and their use is as much to facilitate cropping flexibility and combine-harvesting and to improve the quality of grain as to give yield responses by reducing competition from weeds. The cost of labour, too, is such that the sheer expense of hand-hoeing or mechanical weeding makes chemical weed control essential on cereal crops. Although broad-leaved weeds are now satisfactorily controlled, until quite recently there has been difficulty with annual and perennial grass weeds such as wild oats and black grass, for these are very closely related to the species we use as cereals, and chemicals which destroy the grasses may also be lethal to the crop. New herbicides are, however, now emerging which control these weeds.

*          *          *

The ever-increasing demand for food as populations increase or people become more affluent will reinforce the need for countries in temperate zones to increase cereal production. Farming systems will become more intensive and the mono-cropping of cereals in such systems will increase the risks of pest and disease attack.

Normal cultural methods of pest and disease control are practised by most farmers and these undoubtedly prevent many attacks. Modern crop varieties resulting from intensive work by plant breeders further help to reduce losses.

But when, despite these precautions, the farmer is faced with fields of cereals attacked by pests, diseases, or a combination of these, he will turn to pesticides to protect his livelihood and to ensure a steady flow of cereal grain into the world's larder.

**Further reading**

1  BLAND, B.F. *Crop Production; cereals and legumes.* Academic Press, London and New York (1971).
2  BRITTON, D.K. *Cereals in the United Kingdom: Production, marketing and utilisation.* Pergamon Press, Oxford (1969).
3  FRYER, J. and MAKEPEACE, R. (ed) *Weed control handbook,* 7th Ed. Blackwell Scientific Publications, Oxford (1972).
4  GAIR, R., JENKINS, J.E.E., and LESTER, E. *Cereal pests and diseases.* Farming Press, Ipswich, Suffolk (1972).
5  *Production yearbook.* FAO, Rome (1974).

# 7 Fruit crops: a rather special case

By J. P. Hudson

The world's main field crops fall into three categories: those that are eaten directly by people, those that are eaten by domestic animals, and those that are used as raw materials for industry. Most of the crops in the first category (for example, cereals, potatoes, pulses, etc.) are cooked in some way before they are eaten, but some important exceptions, such as salad vegatables and a wide range of fruits, are eaten raw. It is the fact of being eaten without first being cooked that makes fruit, in particular, a rather special case, because appearance — 'eye appeal' — has become an all-important attribute in the marketing of such products. This has led, in turn, to growers of many fruits being compelled to apply more comprehensive programmes of pest and disease control than has been either economic or necessary for other crops.

Such has not always been the case. No doubt early man found that various wild fruits were both refreshing and pleasant to eat while he was still a food-gatherer, long before crops were domesticated and there was any recognizable system of farming. By a process of selection, often unconscious, improved forms of these fruits were doubtless evolved quite early in man's agricultural history, but the improvement of the perennial tree and bush fruits gained rapid momentum when the art of propagation was developed. This ensured that any gardener who noticed an improved seedling or bud sport (e.g., a mutated branch on a fruit tree with fruits that were better than the existing ones) could deliberately increase the better variety by budding, grafting, cuttings, and a great range of other vegetative methods. In this way improved fruits were perpetuated far beyond the life-span of the original seedling or branch, leading to a highly specialized form of crop husbandry, based primarily on the quality of the product, as manifest by its appearance and taste.

As a result of these developments the world now has a wide range of edible fruits, amounting to thousands of named varieties of several scores of species, that are much appreciated, in great demand, and grown in two entirely different ways. On the one hand is a highly organized fruit growing industry geared to complex systems of storage

Plate 7.1. Low-volume spraying in an apple orchard. The most highly developed systems of fruit growing commonly achieve almost perfect control of all the major pests and diseases, with the result that growers are able to produce high-quality fruit at an acceptable price. (ICI photograph)

and world-wide distribution; on the other hand is a significant production, often of the same fruits, in back gardens or by small-scale farmers, for purely local consumption, using completely unsophisticated methods of production.

## Commercial fruit industry

In round figures, the present global commercial production of fruit[1] is of the order of 25 million tonnes of apples, 7 million tonnes of pears, and 12 million tonnes of other deciduous tree fruits (plums, cherries, etc.), totalling some 44 million tonnes of what can broadly be called 'temperate region' tree fruits. Sub-tropical and tropical fruits grown commercially include 32 million tonnes of citrus (22 million tonnes of oranges) and 40 million tonnes of bananas. About 50 million tonnes of grapes are grown, some for eating but mainly as the raw material for upwards of 30000 million litres of wine.

Although fruits are highly perishable, methods have been evolved that enable most of them to be stored and shipped about the world in a way that would have been impossible before the advent of refrigerated ships. Though commonplace, it is a remarkable feat that fruits like bananas can be made available all the year round, in any part of the world. This is only possible because the fruits can be protected effectively against diseases and pests in the field, in storage, in transit, and in the shop.

With a rising standard of living, especially in industrial countries, there has been a steady increase in the demand for fresh fruits, including relatively expensive exotic species that have to be imported. The demand was greatly stimulated between the world wars by the discovery that fresh fruits provided a spectrum of the vitamins whose dietary importance was then becoming recognized, although most of the world's fruits are probably eaten because they are enjoyed rather than for their nutritional value.

## Unsophisticated fruit growing

The preceding paragraphs refer exclusively to high-technology, commercial fruit growing, but account must also be taken of the entirely different though parallel activity of growing fruits for local consumption, in gardens and small farms, or even semi-wild. These fruits may fall far short of perfection by market standards, and yields may be uncertain and light, but such fruits provide an important source of vitamins for many rural dwellers, particularly in the tropics, where a range of important species, such as mangoes, coconuts, plantains, etc., will virtually grow 'wild', with a minimum of attention.

There are no estimates of the amount or value of fruits grown in this way, but they undoubtedly contribute importantly to the diets of many rural communities in the less developed countries. Families that move into towns are often cut off from supplies of these more or less 'free' fruits, yet they may not be able to afford to buy the much more expensive products of commercial orchards, and this may have adverse effects on their health.

In the present context it is important to distinguish between these two entirely different sources of fruits. Semi-wild and back-garden fruits are usually completely neglected from the point of view of disease and pest control. The plants usually receive no sprays at all, many of the fruits may be blemished (though still edible), and yields may be at levels that would be wholly uneconomic on a commercial scale. On the other hand, well-managed fruit farms currently go to the greatest possible trouble to protect the crop, often to the extent of spraying the trees 15 to 20 times a year with various pesticides. As a result of this expert management, modern orchards are expected to produce heavy regular yields of virtually perfect and unblemished fruits.

This high technology is, however, of relatively recent origin. Early Dutch paintings illustrate the range of luscious fruits that was already available several hundreds of years ago, and no doubt tasted nearly as nice as our modern varieties. The paintings sometimes show blemishes due to recognizable diseases and pests, which were doubtless then accepted as 'natural' to the fruits concerned, and certainly did not hinder the enjoyment of eating them. If anyone considered the blemishes at all, the scabs on apples were probably regarded as essential parts of the fruits, comparable with the calyx and lenticels, while maggots, then believed to be spontaneously generated in flesh, were not connected with inconspicuous free-flying insects such as codling moths.

## Emergence of plant protection techniques

The situation changed very rapidly rather less than a hundred years ago when the nature of fungus diseases was worked out and it was found possible to protect French vineyards against devastating leaf diseases by spraying them with copper compounds. These techniques were taken up rapidly by growers of other types of fruits, who developed a wide range of ingenious equipment for spraying, and the practice of protecting crops by using pesticides was accepted as a new and most important tool in crop husbandry.

Since that time there has been an interesting interplay between three factors in the fruit industry: the ability to protect fruits effectively against diseases and pests; the development of ways of moving big tonnages of fruits in world trade, where it paid to ship only perfect fruits; and the need for local growers in the importing countries to raise the standard of their own produce, to match that of the fruits coming in from abroad. The net result has been that the purchasing public has come to expect that fresh fruits should be entirely free from blemish, and is likely to reject any that are not. Thus, commercial fruit growing has become one of the most highly sophisticated husbandry technologies, leaning heavily on the use of pesticides to ensure that a very high proportion of the product can meet the stringent market requirements for 'perfect' fruit.

It must be admitted that this type of perfection is a somewhat spurious attribute, since it applies mainly to the appearance of the skin, yet fruits are bought to be eaten rather than to be looked at. Indeed, the skin is thrown away in fruits such as bananas, citrus, pears that are canned, and many others. However, there is no sign of any easement of the stringent demands for fruits that are free from blemish, which seems likely to remain a dominant factor in determining the technological level at which market fruit will be produced for the foreseeable future.

## Costs of crop protection

The costs of crop protection in relation to the probable economic benefits that result have been carefully worked out for some pests — as, for example, in the use of soil fumigants against potato cyst eelworm. No such estimates are readily available of the losses in yield and quality of fruits caused by particular diseases and pests, or of the costs of preventing such losses. The calculation of cost-benefit ratios for protecting fruit would be dauntingly difficult, because of the large number of pests and diseases that can occur simultaneously and against which control measures are currently being taken. For instance, in Britain the official agricultural advisory booklet[2] lists 15 pests and 10 diseases of apples which may call for control measures, four and three respectively for pears, seven and three for strawberries, and so on. About 60 different chemicals are currently on sale and 'cleared' for use by commercial fruit growers.

## Fruit crops: a rather special case

The rapidity with which crop protection in fruit gained momentum was partly due to the professionalism of plant pathologists and entomologists. Because of the need for perfection in the product a great deal of research and development has been done on the diseases and pests of fruit relative to the value of that sector of crop husbandry; far more, overall, than on many other crops. This work has been ably assisted by research and development on machinery for spraying, which has resulted in increasingly satisfactory ways of applying sprays. In 75 years spraying has progressed from extremely labour-intensive systems based on headland pumps feeding long hoses and hand-held lances, through tractor-mounted systems applying similar high-volume sprays, to air-blast machines that use less water and cover the orchard much more quickly.

Nevertheless, despite the efficiency of current methods there is no reason to suppose that pesticide application has reached its ultimate peak of effectiveness. New machines now being developed use far less pesticide, applied in much smaller volumes of spray. In the most advanced systems, the machines can be mounted on driverless tractors which follow underground wires that guide them automatically round the orchards. An entirely different but even more promising system is to apply all the chemicals needed in the orchard (pesticides, nutrients, growth regulators, etc.) through a special type of fixed irrigation system that is entirely automatic, sprinkling the trees with the right amount of any selected mixture and dilution of chemicals when wind speeds are very low, as is often the case in the middle of the night.

Chemicals are also used with great effect as herbicides on fruit, though in ways that are quite different from the techniques used in field crops. It is usually difficult, if not impossible, to spray weeds in growing crops without at the same time wetting the crop plants themselves. As a consequence, herbicides can be used in such situations only if they can be applied to the soil before the crop is sown or germinates (pre-emergence), or if they are harmless to the crop plants. This usually means that they are relatively ineffective against at least some of the weed species. With tree crops, on the other hand, it is relatively easy to apply sprays below the crop canopy in such a way that the tree leaves are not wetted, which opens up various interesting possibilities.

The aim in some tree crops is to keep the ground completely free from all plant species, which is easily achieved by spraying beneath the trees with a general plant desiccant. The same type of control can be obtained with bush-type fruits such as vines and blackcurrants, although more precise ways of applying the sprays may be needed to avoid damaging the crops. On the other hand, growers of many temperate fruits now prefer to encourage a ground cover of short-growing plants, either over the whole area or in strips between the rows — in the latter case keeping the ground bare round the bases of the trees themselves. This pattern is easily achieved by using the modern range of herbicides with spraying machines especially adapted for orchard work. Difficulties can, however, still arise from perennial weeds, which are favoured in orchards by the impossibility of cultivating deeply and regularly between the trees. It is nevertheless possible to control most of the perennial weeds by the use of specific herbicides, which are commonly applied direct to troublesome weeds by some form of spot treatment rather than over the orchard as a whole.

## Doubts about the future

From every point of view the current use of pesticides in orchards must be accounted a success story, since the most highly developed systems of fruit growing now commonly achieve virtually perfect control of all main diseases and pests, coupled with precise regulation of the ground cover and control of undesirable weeds. However, as in other sectors, present ideas about crop protection are being questioned in fruit growing. The doubts are partly on grounds of increasing costs, which are very high in comprehensive programmes based on organic chemicals (whose prices are rocketing); but there is a deeper anxiety because of the apparent increase in the resistance of some diseases and pests to the range of agricultural chemicals. It is also increasingly felt that the use of powerful broad-spectrum insecticides over the last 30 years since the highly successful introduction of DDT has created a new situation in which the potential benefit of many natural checks on pests (and perhaps some diseases) has been thrown away.

Legislation has also resulted in a rather new situation, since many countries now exercise extremely close control over the marketing of chemicals for use on crops. For instance, the Agricultural Chemicals Approval Scheme[3] in Britain was concerned with only some 50 chemicals in 1950, yet by 1975 it included nearly 200 different materials in a list that shows little sign of levelling off.

It is said to cost several million pounds to develop a new chemical for agricultural use, which means that the only new pesticides likely to reach the market in the future will be those that are assured of a big sale, such as for use on farm crops grown on a very large scale. This is likely to be to the disadvantage of fruit growers. Several of the earlier pesticides were specifically introduced for use on fruit, but it is unlikely that many such materials will be developed in the future unless they are assured of a much wider application.

Fruit growers may thus be in difficulties if any of the present major components of their spray programmes lose their effectiveness through the development of resistance, unless suitable alternatives happen in the meantime to have been developed which are also useful for some other purposes. Thanks to the ingenuity of chemists a continuous stream of new chemicals will doubtless be synthesized that are more effective, safer, and even cheaper to produce than the ones we already have. But whether or not they will reach the market will depend on the cost of clearing them and the potential market.

All forms of chemical crop protection would be immeasurably easier to apply, and probably more effective, if there were a wider range of truly systemic pesticides. The use of chemicals may also be more rational when there is a better understanding of the way in which molecules pass through the waxy epidermal barrier that protects most leaves from losing water and chemicals under a wide range of weather conditions.

The foregoing remarks apply only to commercial production. Fruit grown in back gardens, on smallholdings, or semi-wild is unlikely ever to receive any effective form of crop protection as long as this requires frequent applications, complex mixtures, or the use of machinery for spraying. The whole situation could change dramatically if effective systemic pesticides were to come on to the market, especially if they could be applied to the soil in granular form and then be taken up by the roots. Such a system of crop protection might be applied on a big scale, if it were simple and cheap enough, and provided it was seen to be leading to better crops.

### Alternative methods of control

In the meantime, the fruit growing industry, and its supporting research and development organizations, are actively seeking alternatives to the present systems of crop protection, which rely more or less completely on chemical pesticides. For instance, there has been much discussion of the 'integrated control' of pests, in which blanket applications of broad-spectrum chemicals are used not as a routine, but only sparingly, in association with other methods of control.

This would call for very careful management, including some form of 'supervised control', in which threshold levels have to be decided for each particular pest below which no control measures are taken. The crop is sprayed at once if infestation reaches some closely defined stage. This system must necessarily be coupled with an accurate method of monitoring, either by visual assessment of the incidence of pests or damage on the trees, or by some other method such as trapping insects by pheromones or lights.

Below the critical level it is assumed that pests are kept in check by some natural means. This leads to the concept of 'biological control' (see Chapter 20). This is obviously an attractive idea. There has been talk of complete biological controls, by which particular pests (or even diseases) are entirely cured or prevented by the use of predators, antagonists, and so on. There have in fact been a few outstanding successes such as the control of prickly pear *(Opuntia* spp*)* in Australia by the caterpillar of the moth *Cactoblastis cactorum,* and of the cocunut moth *(Levuana iridescens)* in Fiji by the Tachinid parasite *Ptychomyia remota.* There have also been more than a hundred other instances where a measure of success has been claimed for some type of biological control, with extremely promising recent results in the case of some glasshouse pests (see Chapter 20), but the method has had rather little success up to now with the major pests of temperate fruits. It is nevertheless by no means impossible that effective systems of biological control may be introduced, even for some of the troublesome diseases like silver leaf fungus *(Stereum purpureum)* of plums, where injecting the tree with a strain of *Trichoderma viride* has given encouraging results.

From time to time there is talk about growing fruit commercially without the use of any chemical sprays. After all, the argument goes, fruit was grown without spraying until a hundred years ago; why not now? The answer must be that the yields from an unsprayed orchard would be hopelessly inadequate, in terms of both quantity and quality, to be economic under present-day market conditions, even though old neglected orchards and back-garden trees do produce a proportion of eatable fruits without any spraying at all.

Moreover, experiments have shown that it is disastrous simply to stop spraying a commercial orchard. The balance of nature is so completely altered by the usual routine of commercial crop protection that a rapid resurgence of serious pests and diseases is likely to make an orchard completely unproductive, at least for some years, if the spray programme is suddenly withdrawn. Until some other reliable method emerges, there is therefore no practical alternative to continuing with the present regime of crop protection if we want high-quality fruits.

There is, of course, always the possibility that some entirely new ways of protecting crops will be discovered. Their development would seem likely to depend upon a far better fundamental understanding of insect and fungal physiology than we have at present. It is a promising sign that more basic work of this type is now being financed, as, for example, at the newly instituted International Centre of Insect Physiology and Ecology at Nairobi.

The final answer to crop protection − the ultimate solution − would presumably be the emergence of crops and varieties that are totally resistant to all the significant diseases and pests. Good progress is being made by fruit breeders towards the production of acceptable varieties that go part of the way towards this objective, but total resistance is unlikely ever to be achieved. In fact, resistance has sometimes worked against crop husbandry rather than for it, because of the rapidity with which some fungi, insects, and mites have developed resistance to pesticides.

*            *            *

The general conclusion must be that fruit growers, and their scientific research workers and advisers, have done rather well in taking quick advantage of all new developments in crop protection. As a result they are able to provide the world with a continuous supply of fruit of very high quality at a price that is acceptable, at least in the more highly developed nations. Their successful fight against diseases and pests relies almost entirely on the skilful use of pesticides and there seems little likelihood that fruit growers will be able to reduce that reliance significantly in the near future − although their technologies will doubtless improve. In the longer term it seems likely that rather more subtle methods may gradually take over, but the day when high-quality fruit can be grown commercially without the use of pesticides still seems a long way off. In the meantime, non-commercial fruit growers at present use few pesticides and probably produce fairly little fruit of marketable quality. The development of cheap, granulated systemic pesticides might give a valuable boost to this sector of fruit growing.

## References

1  ANON. *Fruit: A review.* Commonwealth Secretariat, London
   (1972).
2  ANON. *Fruit growers' guide to the use of chemical sprays.*
   Ministry of Agriculture, Fisheries, and Food, London (1975).
3  ANON. *Approved products for farmers and growers.* MAFF
   Agriculture Chemicals Approval Scheme, London (1975).

Plate 8.1. A striking view of tic beans being attacked by blackfly (*Aphis fabae*), which is a very serious pest of this crop. (ICI photograph)

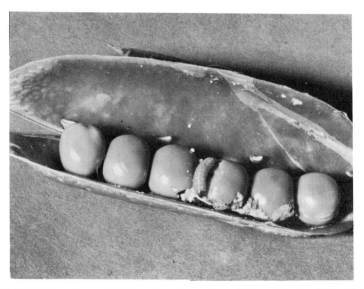

Plate 8.2. Peas in a pod being eaten by a pea moth larva *(Cydia nigricana)*. A heavy infestation of this pest can result in a crop of vining peas being rejected by the buyer. (ICI photograph)

# 8 Vegetables

by Jürgen Kranz

It is more difficult than might be thought to define the term 'vegetable'. Perhaps we can best consider vegetables as plants or parts of plants that are eaten unsweetened, raw, or cooked. Most commonly they are served as salads, soups, or with savouries, like meat and fish. This accommodates vegetables that botanically are fruits (cucumbers, watermelon, tomato, but also avocado pear and breadfruit, etc.) and starchy roots and tubers. Potatoes, sweet potatoes, cassava, and yams may not be accepted by everybody in this context, for they can assume the rank of staple foods. We shall include them here and leave it to the specialists to argue about classification. Pulses, however, traditionally are kept apart.

## Vegetables in the human diet

Vegetables have a vital nutritional role to play. Essentially they provide vitamins and minerals for man. This is true particularly for vitamins C, K, and P (provitamin A), as well as for some of the B vitamins. Among the minerals, calcium, iron, magnesium, and potassium are most important. Some vegetables also supply significant amounts of energy and proteins. Vegetables thus have a key position in averting malnutrition, diseases, and disorders in man.

Turning from necessity to luxury, culinary art rests largely on the variety of flavour and digestive values that vegetables contribute to our dishes.[5] Most *hautes cuisines* provide for special vegetable courses. Non-starchy vegetables thus '. . . raise . . . eating from the level of a mere satisfaction of hunger and of our nutritional requirements to that of a pleasurable occupation.'[4] Chinese as well as most other oriental cookery is unthinkable without vegetables.

Vegetables were eaten in early civilizations and cabbages, tomatoes, chillies, and onions have become essential items of traditional dishes. National dishes are often characterized by particular vegetables. Irish stew, ratatouille, imam bayilda, letscho, sauerkraut, borsch, and many others have even gained international reputations. There are, however, large differences in vegetable consumption between countries.

The Italians, for instance, eat twice as much vegetables as do the Dutch (170 kg compared with 81 kg/*per caput* per annum).

## World production and export of vegetables

Commercial production of vegetables for sale and for export is only a few centuries old. In some regions it has become a major occupation and an important foreign exchange earner. Selected production figures are given in Table 8.1. Figures like these do under-state the real volume of production. They do not include vegetables grown in private gardens nor a proportion of those traded in local markets. Nevertheless, they seem to be fairly representative as indicating a slight gross rise in production due to higher demands. This increase is due to both greater acreage and higher yields per hectare. It is noteworthy that North America and Europe produce much more than Africa and Asia. The trend (not shown in Table 8.1) in more industrialized countries is especially to increase production of salad crops (tomatoes, lettuce, and cucumbers) grown mainly under glass.

The figures in Table 8.1 show that the increase in production is hardly enough to improve significantly the food supplies for a growing world population. Temperate-zone vegetables are particularly insufficient in the tropics, where they are becoming more and more popular, owing partly to the influence of health authorities. Destruction of surplus vegetables in the European Common Market is by no means representative of the situation in other countries.

Vegetable production is becoming more market-orientated in order to ensure a fair return on capital and labour input. The annual figures are of the order of 10 million tonnes. Owing to their high water content and soft structures, vegetables are not easy to maintain in a marketable condition if transported over long distances, and their shelf life may suffer. Most exported vegetables are therefore processed or preserved or merely go to neighbouring countries, as in Europe. In Europe, only Italy, the Netherlands, and Portugal are completely self-sufficient in vegetables. Tomato with its various products (whole fruits, juice, purée, etc.) is outstanding as a most important vegetable for processing, followed by asparagus, green beans, peas, and sweetcorn.

## Some difficulties

Except for wild or semi-wild plants that are collected locally for food, most vegetable species are cosmopolitan and specialized for high yields

Table 8.1. *Selected production figures for vegetables in 1972, compared with 1961-5*
*All values in 000 tonnes per annum*

| | Africa | | America (N. and Central) | | America (South) | | Asia | | Europe | |
|---|---|---|---|---|---|---|---|---|---|---|
| | 1961-5 | 1972 | 1961-5 | 1972 | 1961-5 | 1972 | 1961-5 | 1972 | 1961-5 | 1972 |
| Artichoke | 41 | 40 | 24 | 32 | 20 | 74 | 24 | 11 | 663 | 1 127 |
| Bean (fresh) | 153 | 187 | 861 | 947 | 89 | 114 | 397 | 599 | 1 004 | 1 263 |
| Cassava | 31 948 | 46 220 | 538 | 713 | 25 217 | 36 168 | 18 364 | 22 188 | — | — |
| Carrot | 165 | 190 | 922 | 1 098 | 167 | 180 | 447 | 789 | 2 443 | 2 998 |
| Cauliflower | 106 | 128 | 131 | 154 | 48 | 58 | 564 | 770 | 1 969 | 2 195 |
| Cabbage | 351 | 436 | 1 308 | 1 386 | 170 | 208 | 4 308 | 5 559 | 7 645 | 7 645 |
| Chilli | 539 | 748 | 313 | 479 | 84 | 97 | 385 | 653 | 1 192 | 1 436 |
| Cucumber | 184 | 181 | 654 | 822 | 7 | 14 | 1 291 | 2 060 | 1 210 | 1 639 |
| Eggplant | 176 | 269 | 33 | 41 | 1 | 2 | 1 333 | 1 729 | 394 | 479 |
| Melon | 1 846 | 2 195 | 2 490 | 2 265 | 722 | 746 | 5 794 | 7 565 | 3 245 | 3 817 |
| Onion | 1 042 | 1 018 | 1 421 | 1 645 | 845 | 997 | 5 562 | 7 349 | 3 151 | 3 742 |
| Pea (fresh) | 80 | 103 | 1 280 | 1 261 | 89 | 125 | 321 | 374 | 1 482 | 1 925 |
| Potato | 2 022 | 2 819 | 14 990 | 15 849 | 7 010 | 7 997 | 38 129 | 46 227 | 138 283 | 128 133 |
| Pumpkin | 435 | 591 | 75 | 60 | 489 | 464 | 1 108 | 1 096 | 778 | 704 |
| Sweet potato | 4 628 | 5 839 | 1 314 | 1 266 | 2 399 | 3 161 | 99 137 | 114 366 | 151 | 83 |
| Tomato | 2 015 | 3 054 | 6 139 | 7 727 | 1 181 | 1 681 | 3 091 | 5 227 | 8 207 | 11 412 |
| Yam | 18 194 | 18 858 | 122 | 197 | 71 | 97 | 42 | 31 | — | — |

Source: *FAO production yearbook 1972.*

and high standards of quality. This, together with great genetic uniformity, may make them highly vulnerable to diseases, pests, and weeds.

Modern commercial vegetable production usually requires large capital and labour inputs, refined skill, and good marketing facilities. Consequently production costs per unit of product are high. On the other hand, demand for vegetables can be remarkably inelastic, limited only by high prices and the human capacity (or desire) to eat them. Often the structures of wholesale and retail markets accentuate variations in financial returns from vegetable production. The variations can be extremely sharp, entailing considerable risks for the growers.

There may still be scope for increased productivity. More fertilizer, better irrigation facilities, better varieties, and improved agricultural and glasshouse technology could help. Unfortunately, higher productivity per hectare resulting from increased inputs gives rise to more and increasingly severe diseases, more pests and perhaps weeds, and, consequently, to the need for increased plant protection.

Soft and juicy vegetables are particularly exposed to post-harvest diseases and pests, which often cause severe losses. In the tropics, this toll is particularly high because of poor transport systems, lack of cold storage, and inferior marketing facilities, as well as to delayed harvest. Considerable effort is required to keep stored vegetables healthy until they are eaten. Processing could obviate a large proportion of post-harvest losses everywhere, and vegetable processing is in fact becoming more important. This may take the form of canning, quick freezing, dehydration, ionizing radiation, extracting vegetable juice, or preparing ready-made dishes, but it always requires impeccable quality. High quality is certainly not feasible without plant protection, which tends to be included in the production contract commonly used by processers.

## Losses from diseases, pests, and weeds

Fungi, viruses, and bacteria as pathogens; insects, nematodes, rodents, snails, and other animal pests; and, of course, weeds, all reduce yields and may affect the quality of the unprotected crop.

Insects or rodents may eat, and a fungus may rot away parts of the roots, leaves, stems, or fruits meant for human consumption. Such damage is obvious. But if an insect merely tastes and then abandons a tomato fruit, or a fungus causes a single small dark or corky spot, the crop may fail to reach the high standards required by housewives and vegetable processers and become a complete loss. Because of this, without crop protection farmers would have to accept tremendous losses, when grading their produce, added to the pre-harvest losses already suffered. All this is by no means confined to affluent societies where appearance often counts more than taste. Often, though, losses may be less obvious: sucking insects, nematodes, or grubs in the soil; diseases on non-harvestable plant organs; and weed competition may have reduced growth or fruitset, delayed ripening, caused malformations, and so on. At any given level of production — where no further advances in productivity can be expected — all these losses effectively reduce the already limited potential available to feed the world population.

Table 8.2 summarizes estimated losses in yield in some selected

TABLE 8.2. Estimated losses in some selected vegetables owing to diseases, pests, and weeds*

| | Losses as percentage of potential yield† due to | | | Total losses |
|---|---|---|---|---|
| | diseases | pests | weeds | |
| Asparagus ‡ | 7 | 12 | — | 19 |
| Cabbage | 4 | 9 | 5 | 18 |
| Carrot‡ | 7 | 2 | — | 9 |
| Cassava | 16 | 7 | 9 | 32 |
| Cauliflower | 3 | 5 | 5 | 13 |
| Cucumber‡ | 13 | 9 | 4 | 26 |
| Chilli-pepper‡ | 12 | 6 | 6 | 24 |
| Onion | 13 | 10 | 8 | 31 |
| Pea (fresh) | 7 | 4 | 23 | 34 |
| Phaseolus bean (fresh) | 8 | 8 | 10 | 26 |
| Potato | 22 | 7 | 4 | 33 |
| Salad ‡ | 10 | 6 | 6 | 22 |
| Sweet-potato | 5 | 9 | 11 | 25 |
| Tomato | 12 | 7 | 5 | 24 |

* Compiled from estimates in ref. (2).
† Potential yield—yield which would have been harvested without diseases, pests and weeds.
‡ Figures from the USA only. All the others also comprise Europe (without the USSR), Africa, Asia (without China), Oceania, and South America.

vegetables. Such estimates are crude, but may serve to indicate global magnitudes. These figures are not, however, based only on unprotected crops. Insufficiently protected crops also contribute to them. And it is quite likely that they do not always reflect losses due to grading. Overall figures, as in Table 8.2, often disguise the effect of an attack at the individual grower's level. Averages tend to smooth out fluctuations between regions and years. An example may elucidate this point. On cauliflower, insects cause an average loss of about 3 per cent in Britain. Local losses of this crop due to one pest alone, however, may be as high as 89 per cent.[2] Such variation is not at all uncommon. Even smaller average losses, if very likely to occur, may bring plant protection into action.

The data in Table 8.2 do not give any clue to effects on quality. These are largely determined by adverse changes in the composition of the vegetable, or new compounds formed either by the plant itself or by the pathogen or insect. Mycotoxins are a dreaded example.

Protection of vegetables from losses is more commonly justified by experiences of the following kind. A leaf-spot disease of tomatoes in the U.S.A. may severely reduce yield of marketable fruits; in unprotected plots fertilizers raised this yield only from 18.8 to 40t/ha; when a copper fungicide was applied as well, yields of 50-60 t/ha were harvested. Plant protection generally increases yields between 10 and 100 per cent compared with untreated crops, though higher increases are by no means rare. In Indonesia, for instance, a nearly threefold increase in yield of potatoes could be achieved by seven applications of a fungicide against late blight.

Hardly any vegetable grower can afford to risk severe damage, but neither can he squander input. Thus, a cost-benefit ratio is involved. If the benefit is known to be high, growers will more readily use chemical plant protection – often as insurance. Thus, in Britain, only 2 per cent of the onion fields and 10 per cent of the celery, but 197 per cent (i.e., repeated spraying) of the acreage of carrots are regularly treated with insecticides.[9]

Such losses affect the consumer in the first place through vegetables becoming unobtainable or through soaring prices. The grower also suffers. More often than not there is no compensation for him for higher prices for less produce – the middlemen usually takes that profit. In brief, losses in production mean losses in income for the individual producer as well as for the community, smaller turnover, and less tax collected. There can also be other repercussions.

## Protecting vegetables from losses

As we have seen, the protection of vegetables is essential in any horticultural practice, whether in commercial production or private gardens. Scarecrows, mole traps, slippery bands on tree trunks, cutting and burning of infested plants or plant parts are commonplace even for amateur gardeners. Weeding is essential. Classical authors, like Pliny, listed a variety of remedies against diseases and pests, including prayers and sacrifices. But they also recommended using such materials as salt, nitrate, amurca (olive dregs), ashes, vinegar, and urine. A good amateur gardener also knows that not every kind of vegetable does well in his garden; and he is aware that certain cultivars of a particular species are more resistant to diseases or tolerant to pests than others. Some natural crop resistance, even if only moderate, is a *sine qua non* for efficient chemical plant protection. Resistance is often the only economic control available for virus and bacterial diseases.

Sowing time, spacing, proper manuring or fertilizing, rotation, and many other things will have to be observed in order to harvest good crops of big potatoes, cabbages, carrots, and beans. When the gardener buys his seed he takes care that it is not too old and that a seed dressing is applied; or he makes sure that the seed has been certified free from disease, grown in a disease-free area, or cleared of viruses by heat treatment or meristem culture. And he tills his ground carefully, removing or deeply burying all litter. He also eradicates all volunteer plants quickly. Why does he do this? Partly to secure optimal growth conditions, but a great many of these established practices also prevent diseases and pests from surviving the winter, and then from meeting their host plants in susceptible stages, or from spreading rapidly within the crop. All this is often unintentional plant protection, but it is also regularly employed in commercial vegetable production where appropriate.

The greengrocers who cater for the majority of town dwellers get their supplies from commercial vegetable growers. These growers cannot afford bad years for they have to recover investments as well as running costs, and to earn at least a small income. This usually means more than one crop per season, increased input, more high-yielding cultivars, improved technology, and, eventually, glasshouses for greater independence from the weather and seasons. All these measures increase the liability to losses from diseases, pests, and weeds.

Plant protection therefore becomes more important, more sophist-icated, and more intensified than in a private garden. Modern chemical compounds may be indispensable to prevent losses. Soil disinfection by chemicals or steam is routine in glasshouses and plots repeatedly cropped with the same crop. It is mostly directed against fungi, which kill the seedlings and thus prevent establishment of the crop, and against nematodes. Seed dressing is done for the same reasons. These chemical compounds usually dissipate quickly, leaving no harmful residues; or the doses applied are so small that even with dangerous compounds like the mercurials, the accumulation is below the natural level of contamination from rain. If uptake by plants occurs, these compounds are usually metabolized and made harmless.

The need to spray chemicals becomes more urgent with increased intensity of production, with increasing wholesale prices of the crop, with higher temperature and, for diseases, with increased humidity. For instance, in the tropics tomatoes can often be grown only with frequent applications of fungicides. Up to 20 spray rounds (twice a week) are considered necessary to protect the celery crop against a leaf-spot disease in Florida. Although such treatment is costly, farmers still make a profit without causing hazards to customers. Hand weeding is rarely economic in commercial vegetable production. Shortage of labour then forces growers to substitute herbicides, which for vegetables are usually applied weeks ahead of planting.

Hazards are generally more imminent when pests are to be controlled. Most insecticides permitted for use on vegetables are harmless at the prescribed doses. But there is a snag: nearly all vegetables are endangered by diseases and pests right up to harvest time. Growers may thus feel tempted to provide protection up to this point, disregarding the makers' instructions and consequently endangering consumers. As this is beyond the control of the makers and difficult for local authorities to control, there is a trend towards insecticides with short persistence. Today, a safe harvest of tomatoes and cucumbers is possible as soon as three to four days after the last application of some insecticides in the field. Others require a waiting period from seven to 60, or even 120 days, depending on the crop, dose, climate, etc.

Unfortunately, some of the insecticides with short persistence have a high acute toxicity for man. Longer-lasting but less toxic compounds still have their place for the control of pests attacking a crop early. Minimizing the hazard to consumers and the environment is an obligation for research workers, pesticide manufacturers, growers, and consumers alike. Growers must be extremely conscious of what they are doing. Customers should be less meticulous about the appearance of vegetables, thus helping the grower to accept the idea of tolerable pest population or disease levels, and thus reduce spraying. Manufacturers and research workers have to continue with their research on side effects.

The study of epidemiology (population dynamics) could certainly help to optimize control of diseases and pests, by emphasizing disease and pest management. Improved knowledge of population dynamics is basic also to the application of newer developments which seem to have less undesirable side effects: for example, chemosterilants, sexual attractants (*pheromones*), juvenile hormones, synergistic compounds, insecticides from natural sources, and skimmed milk against vectors of a few viruses, etc. Some have shown promise in the protection of vegetables (see Chapter 20). New application techniques such as the use of granules, capsules, etc. could also make insecticides less hazardous in the future.

Protection of the environment does not normally arise in glasshouses, but it does become important where vegetables are grown outdoors in rotation, or as secondary crops in suitable climates, as after rice in some regions in the Far East. Here all the implications of pesticide application on other biocomponents of the site apply, such as the effect on harmless or beneficial insects and on fish, the induction of resistance to pesticides amongst pests, and changes in plant communities.

There is little scope for immediate biological control in short-lived vegetables. The egg parasites or the virus diseases of insects are perhaps exceptions. *Bacillus thuringiensis*, which is produced commercially and is effective against some lepidopterous caterpillars, may be of use (see Chapter 20). Predators, parasites, and diseases usually lag behind, and hardly have time to catch up with the pests before harvest. Nevertheless, they deserve protection in view of the long range and stabilizing effects which they may have. Neither research nor advisory work is so far proving very productive in these matters, particularly in the less developed countries. Such research is very costly and the richer countries have to carry most of this burden, as, indeed, the U.S.A. and other countries are doing. However, politicians and the public can find it difficult to grant funds because of surpluses of food that are produced in some of the developed countries.

*        *        *

To summarize, vegetables are essential for human nutrition and well-being, as well as being an important source of income to producers.

*Vegetables*

Comparatively high input and skill are usually required for their production. Most vegetables have soft tissues and a high water content throughout their lives, and for this reason they are often highly susceptible to diseases and insect attacks. Intensification of production, necessary to meet rising demand, is an aggravating factor. For instance, for the potato crop alone, about 300 diseases, in addition to pests, have been recorded in the world. About 30 per cent of the potential yield of this crop and of many other vegetables fall victim to diseases, pests, and weeds every year if they are left unprotected. Plant protection has therefore become an integral part of vegetable production, and is still increasing in importance. It is an illusion to imagine, as it is sometimes suggested, that we can do without any of the available weapons, such as pesticides. This would have drastic detrimental consequences for mankind. We must, however, wield these weapons wisely and still more efficiently.

## References

1. ANON. *Pest control. Strategies for the future.* National Academy of Sciences, Washington D.C. (1972).
2. CRAMER, H.H. Plant protection and world crop production. *PflSchutz-Nachr, Bayer* **20**, 1-523 (1967).
3. CHUPP, C. and SHERF, A.F. *Vegetable diseases and their control.* Ronald Press, New York (1960).
4. DUCKWORTH, R.B. *Fruit and vegetables.* Pergamon Press, Oxford (1966).
5. ESCOFFIER, G.A.A. *Guide to modern cookery.* Heinemann, London (1959).
6. *FAO Production Yearbook 1972.* FAO Rome (1973).
7. RABB, R.L. and GUTHRIE, F.E. *Concepts of pest management.* North Carolina State University Press, Raleigh N.C. (1970).
8. SHURTLEFF, M.C. *How to control plant diseases in home and garden.* 2nd ed. Iowa State University Press, Ames (1966).
9. STRICKLAND, A.H. Pest control and productivity in British agriculture. *Jl R. Soc. Arts,* **113**, 62–81 (1965).
10. TINDALL, H.D. *Commercial vegetable growing.* Oxford University Press (1968).

# 9 Pest control in livestock production

by D.E. Jacobs

The process of evolution has ensured that there are few sites on earth that do not harbour a variety of forms of life. A sand-dune, a ditch, a rotting tree-trunk — each swarms with a characteristic range of plant and animal species adapted to that particular environment. If one of these species becomes 'noxious, destructive or troublesome to man, or to his interests, it is branded as a pest.[2]

The pressure for species to diversify ensures that every available source of nutrition and shelter is utilized, and nature has not exempted the use of living things as habitats. A fruit tree, for instance, sustains organisms as varied as fungi, nematodes, and arthropods. Some of these are totally dependent on the tree for survival and procreation, while others are more casual visitors. Some are beneficial or neutral, others are harmful to the tree. The higher mammals, including man himself and domestic livestock, have been termed 'mobile ecosystems' for each is host to an abundant flora and fauna. A louse running through the fleece of a sheep is not very different in some respects from a beetle scurrying across the lawn, and pesticides employed for plant protection are often used to protect the health of animals.

Just as a tree provides not one but a variety of habitats — the root, wood, bark, and leaves — so different parts of the animal body have their characteristic passengers. Many have discovered an external surface with environmental characteristics rather different from the skin: the alimentary canal. For, as a tube with an opening at each end (the mouth and the anus), the contents of the digestive tract are, physiologically, outside the animal. Thus, under certain carefully controlled conditions, it is possible to apply selected pesticides in the control of gastro-intestinal parasites without endangering the animal host, as in the case of bot-fly larvae *(Gastrophilus* spp*)* in the stomach of the horse.

In their search for a habitat, some parasitic species have progressed beyond the external surfaces of the animal and have penetrated the

Plate 9.1. Spraying against cattle tick in Kenya; a thorough soaking with the spray is necessary if treatment is to be effective. Recent years have seen considerable advances in the development of veterinary insecticides and acaricides as well as in pharmaceuticals for treating internal parasites. (Shell photograph)

tissues. Here the association between the host and parasite is intimate and control by chemicals necessarily involves the introduction of toxicants into the body. Most of these toxicants are not called pesticides, although they have the same purpose; they are dealt with by the pharmacologists.

### Host-parasite-environment interaction

There is a fundamental difference between a living habitat and an inanimate niche: the former can actively respond to the presence of the life-forms that have adopted it as home. Animals set up physical, chemical, and enzymic barriers to prevent the establishment of unwanted populations, they respond by scratching, rubbing, and licking, or they react either with non-specific cellular processes or with target-orientated immunological ones to contain and destroy invaders. These mechanisms work best when the host is healthy. A low level of nutrition or any other form of environmental stress will impair their effectiveness.

Parasitic species in their turn have evolved methods of evading the host's defences; for example, one trematode (*Schistosoma*) disguises itself by incorporating host protein on to its cuticle so that it cannot be recognized as 'not-self' by its adopted provider. Thus, parasitism is not the passive acceptance of an unwanted invader but a complex dynamic process involving the interaction of three major factors: the host, the parasite, and the environment.

In the case of livestock, two of these factors, the host and the environment, have been radically changed since the progress of domestication began some 9000 years ago. From early times man has used domestic animals to provide food and manure, clothing and leather, transport and power, employment and protection, company and recreation. Today, even the by-products of the abattoir have innumerable applications in the preparation of articles as diverse as plywood and steel, photographic paper and fire extinguishers, furniture and textiles, wine and soap, pharmaceuticals, and much else besides.

In some instances man has adapted his life-style to fit in with that of his livestock; thus the Lapps follow their reindeer herds. More often, however, man has imposed his will on the beast and in the process has restructured anatomical, physiological, and behavioural characteristics: compare, for example, a Landrace pig and a wild boar. As well as manipulating the genetic qualities of his stock, man has altered the environment in which his animals are reared. Today's industrial pig unit, for example, with hundreds or thousands of animals kept on a small area of concrete, contrasts with the feral sow foraging in her several hectares of woodland accompanied by the few piglets she manages to suckle. Even in regions where extensive husbandry methods are practised, there is a trend towards the replacement of natural pastures with improved forage plants and the use of irrigation, fertilizers, and crop protection.

### Pests, parasites, and productivity

Man tries to increase livestock productivity by breeding superior animals and by controlling the factors that prevent them from fulfilling their maximum potential. Success in these ways is necessary in order to meet demands for high-quality protein and to satisfy the meat requirements of the more prosperous countries. Milk, eggs, wool, and leather continue to be animal products of major economic importance.

The period over which these genetic and husbandry changes have taken place is tiny in evolutionary terms. The warble fly managed to outlive its host of the Oligocene period, which is now extinct, but this adaptation took place during a 40-million-year interval, whereas a commercially acceptable pig can be produced from the wild boar in as little as five generations. The rapid changes now in progress inevitably disturb any equilibrium that may have developed between host and parasite. This may be to the detriment of the latter but more often parasite numbers, and hence the risk of disease, increase. Crowding animals increases the opportunity for contamination and transmission of infection; modern production methods sometimes cause stress in the host with a lowering of resistance, and more susceptible breeds may be brought in to replace lower-yielding native stock.

A healthy animal will generally be able to tolerate a few parasites without ill-effect, but larger numbers can cause overt disease, or even death. Between these extremes is the state known as 'less than optimum productivity',[6] where parasite populations of intermediate size reduce the economic performance of livestock without death or obvious illness. The animal may grow more slowly, or require more food to fulfil its potential. Alternatively, milk, egg, or wool production may be reduced in quantity or quality. Such subclinical effects are often unnoticed or unsuspected; they rob mankind of more animal protein than many of the more dramatic diseases that are readily recognizable.

With new methods of husbandry, patterns of disease alter: new parasites become prominent, the symptoms of familiar diseases may change, and their economic significance wax or wane. Twenty years ago, coccidiosis of chickens was primarily an acute disease of chicks characterized by heavy mortality and caused by a single species, *Eimeria tenella.* Now with the development of the modern broiler industry at least five coccidian species have been incriminated in a disease complex manifested mainly by reduced productivity.[7] The annual expenditure on coccidiostats to prevent these losses amounts to some £50 million.

## Pests, parasites, and nutrition

The deleterious effects of parasites are increased if the host's ability to combat infection or to compensate for tissue damage is impaired by poor nutrition or other predisposing factors. An example is the 'thin sow syndrome' seen in Britain some years ago. The precipitating factor seemed to be the feeding of groups of pigs with restricted diets; in these groups the less thrusting individuals were bullied, failed to secure their full share, and instead filled their stomachs with bedding material.[11] Large numbers of nematode larvae were thus ingested and these grew to maturity in the alimentary tract, interfering with digestion and thereby increasing the food requirement of the already deficient sows. Thus, a downward spiral was established and these infected animals required yet more food to maintain body heat once they had started to lose their insulating layer of back-fat.

The 'thin sow syndrome' occurred as a result of economic pressures in a highly developed pig industry, but undesirable host-parasite-environment interactions of this nature are particularly prevalent where climatic or social conditions allow only a marginal level of nutrition for the livestock population. It is often difficult to determine whether a heavy parasite burden is the cause or the result of poor bodily condition.

Even in the absence of the insidious effects of parasitism, livestock cannot thrive if they are undernourished; and as the number of animals increases to satisfy the demand for protein, the animals' food supply has to be increased. Any factor affecting the abundance of animal feedstuffs may have immediate and far-reaching effects on livestock production. This is illustrated by the devastating effect of the 1972-3 drought on livestock in the Sahelian zone of Africa. Again, a shortfall in the 1972 Russian harvest necessitated the importation of large quantities of grain for livestock feeding, influencing world prices. Price fluctuations on the world cereal market exaggerate the normal cyclical patterns of supply and demand in intensive pig-rearing areas. Because of an increase in cereal prices, the number of young animals kept for breeding in the EEC fell by 6½ per cent during the year ending April 1975. Similarly, a sharp decline in the United States pig herd followed the corn leaf blight epidemic of 1970. In this way, pests and parasites of cultivated plants become relevant to this chapter because of their effect on crop yield.

### Economic parasitology

*Vectors*

Host-parasite-environment interactions can become particularly complex when more than two animals are involved. Ticks, for example, are eight-legged arthropod parasites that may use several different hosts during their life-cycle as well as spending several months of each year on the ground. Moreover, they themselves are frequently parasitized by one or more of a long list of bacterial, rickettsial, viral, and protozoal organisms, many of which can produce disease when transmitted to livestock. Babesiosis, for example, is a protozoal disease, characterized by fever and haemolytic anaemia, that became disseminated to susceptible populations round the world with tick-infested cattle in the late nineteenth centrury.[9] In Australia there were reports in 1897 of outbreaks with mortality rates of 30 to 50 per cent; 4000 of 6000 head were lost on one occasion. Losses declined as a more stable epidemiological pattern emerged and more effective control measures were evolved. *Babesia bigemina* and the carrier tick *Boophilus annulatus* were finally eradicated from the U.S.A., where annual losses had risen to $ 40 million, after some 40 years. Even in 1974, after the development first of the organochlorine and, subsequently, the organophosphorus, carbamate, and phenamidine acaricides, as well as with much scientific information on the biology of the tick, the total annual losses due to babesiosis and the other detrimental effects of ticks in the state of Queensland were estimated at $ 50 million. [10]

Many other diseases of veterinary importance are spread by invertebrate pests. One of these, trypanosomiasis, spread by the tsetse fly, has been described in Chapter 3. A sevenfold increase in the cattle population of seven million square kilometres of Africa, equivalent to 1.5 million tonnes of additional meat each year, could be achieved with the elimination of this condition. [4]

A vector does not necessarily have to feed on the animal it damages. Certain snails, for example, act as intermediate hosts for schistosomes (also mentioned in Chapter 3), some species of which infect cattle, and the liver fluke *(Fasciola hepatica)*. In the latter case, parasites leaving the snail *(Lymnaea* spp*)* encyst on herbage and wait to be eaten by a suitable mammalian host. The liver damage produced by migrating immature flukes can cause sudden death: 80 000 ewes from a population of 1.5 million sheep were lost in North Wales during the winter of 1958-9. The adult flukes cause chronic wasting and anaemia but are also responsible for 'less than optimum productivity' in beef and dairy cattle grazing affected pastures. The cost of fluke infections in the United Kingdom has been estimated at £50 million in years of average incidence.[5]

### Blood-suckers, flesh-eaters, and nuisance flies

Many of the subclinical effects of fascioliasis are attributed to the fact that the adult fluke feeds on the blood of its host. Similar debilitating effects are produced by other pests including ticks, which withdraw up to 3 ml of blood each when they engorge. Sometimes the saliva of the tick also contains a toxin that may cause paralysis and death from respiratory failure.

The small wounds inflicted by feeding ticks irritate the animal, causing 'tick-worry', and they may become infected with bacteria. This reduces the value of the hide for leather production, costing $ 3.1 million annually in Australia alone. Another pest of concern to the tanning industry is *Demodex*, a microscopic mite related to groups that cause diseases such as scabies in man and sheep scab. Of greatest importance to the leather trade in the northern hemisphere, however, is the warble fly *(Hypoderma* spp*)*. This insect deposits its eggs on the hairs of cattle; the hatched larvae bore into the body tissues and after a migratory period of several months come to rest under the skin of the back. They make a breathing hole, through which they eventually escape having reached a length of some 2.5 cm. In Britain, up to 55 per cent of the hides from certain areas are affected in peak months, with an overall incidence of 8 to 14 per cent.[1] In contrast, the corresponding figure for Eire in 1974 was only 0.05 per cent because of the country's compulsory eradication policy utilizing systemic insecticides. Hide damage, together with an adverse effect on milk and meat production in affected herds, is thought to be responsible for losses of £4.5 million per annum in Britain.

When, as in the case of the warble fly, dipterous larvae invade and feed on the tissues of living vertebrates, the term 'myiasis' is used. A most unpleasant form of cutaneous myiasis known as blowfly strike occurs when egg-laying calliphorine flies *(Lucilia* spp*)* are attracted to skin soiled with faecal material or to wounds, which may be caused by the bites of nuisance flies or ticks. The active maggots hatching from the eggs devour the living flesh, excavating large open sores. In 1962-3, the cost of blowfly strike to the Australian sheep industry was estimated at £9.9 millions.[3] In the U.S.A., the related screw-worm problem (cutaneous myiasis caused by *Cochliomyia hominivorax)* was of such a magnitude that a biological control programme was implemented, with great success, involving the production and release of several thousand million male flies sterilized by irradiation (see Chapter 20).

Livestock in all parts of the world are plagued by the fleeting visits of many species of insects that come not to lay eggs but to obtain a protein meal. These include midges, simuliids (black flies), psychodids (owl-midges and sand flies), mosquitoes, tabanids (horse flies), house-flies, and stable flies. They damage the skin and expose it to secondary infection and myiasis, while some also suck blood or transmit other disease organisms. The irritation they produce agitates feeding animals, reducing the time spent grazing and lowering the efficiency of feed conversion and milk production. Their salivary secretions sometimes induce hypersensitivity reactions such as the seasonal dermatitis of horses caused by *Culicoides.*

Cattle become very nervous when warble flies are on the wing and Taylor[12] satirically relates the consequences of an incident where a group of valuable heifers was stampeded over a cliff. He takes the opportunity of reminding us that many of the statistics on 'economic parasitology' are extrapolations of inadequate data and are to be interpreted with due care. However, the foregoing paragraphs give at least some indication of the 'cost to the international larder' produced by vector-borne diseases and the activities of arthropods.

*Parasitic gastro-enteritis*

Production losses can be very difficult to quantify in financial terms. In theory, the sudden elimination of all internal helminth parasites affecting milk yield could cause a glut of milk in some areas without a proportionate increase in income; this would be followed by a compensatory reduction in the size of the dairy herd, which in turn could release grassland, grain, and protein supplement for other purposes. Thus, the benefits of parasite control in this case cannot be equated with the value of the extra milk produced, but rather with the gains accruing as a result of the alternative use of the liberated resources. Again, quicker fattening of pigs would not only save feed but allow a faster turnover of stock. This would create a need for more breeding animals, in turn demanding greater capital investment. The situation is complicated still further when the final amount of loss in monetary terms depends on the managerial skill of the farmer in choosing between various remedial actions. Computer modelling systems are being developed to investigate the financial complexities of disease control so that realistic cost-benefit estimates related to total production systems can be derived.

However they are calculated, the losses caused by internal parasites are substantial, estimated at $330 million a year by the U.S. Department of Agriculture during the 1950s and £12.9 million for sheep in Australia.[3] Elimination of the parasite population does not inevitably result in increased productivity, however; Gordon[6] describes a long-term trial in which he demonstrated gains in only 7 of 13 years. The variations in response can often be attributed to the effect of the weather on the preparasitic stages of the helminth; disease forecasting systems are being developed to enable the nature and intensity of prophylactic measures to be predetermined for each period of risk.

Consideration of gastro-intestinal nematodes takes us to the interface between the pesticide and pharmaceutical industries. Of the two modern broad spectrum anthelmintics in most frequent use at the time of writing (1975), one, thiabendazole, is also employed as an agricultural fungicide whereas the other, tetramisole, enhances certain immune responses and may lead to advances in the treatment of some malignant, infectious and immune deficiency diseases.

## Pesticides and animal productivity

The elimination of human disease is a primary objective in its own right, but with diseases of livestock the primary concern must be the enhancement of economic productivity, although this aim is generally consistent with humanitarian considerations. An animal health programme must not cost more than the potential value of the benefits to be derived from its implementation. It often happens that disease control objectives must be subordinated to wider agricultural strategies, the degree of compromise depending upon the impact of the disease— few cattle will be raised in tsetse regions until the problem of trypanosomiasis is resolved. On the other hand, an irrigation scheme may be carried through with an enhanced risk of bovine schistosomiasis if, through an improvement in the quantity and quality of available feedstuffs, there is a net benefit to the welfare of man and his animals.

Adequate food supplies are essential for optimum livestock production. In this context Potter[8] has stated that while improved varieties and better cultural practice will provide, or have provided, the potential for efficient crop production, this often cannot be realized without adequate control of pests such as plant parasitic arthropods, fungi, and nematodes. The elimination of weeds can also be significant. In Britain, for example, an effective herbicide for killing bracken will release for more intensive grazing some 200 000 hectares of upland pasture that cannot be cultivated for climatic, topographical, or economic reasons. In Africa, the use of arboricides is contributing to the clearance of scrub which at present reduces the grazing potential of 10 million hectares in the semi-arid zones of Kenya. After harvesting, stored crops must be protected from rodents and weevils to obtain their full benefit.

The concept of host-parasite-environment interaction implies that the pest is only one component of clinical or subclinical parasitic disease, and modern control measures take account of this fact. Thus, there are a number of possible ways in which the host-parasite balance can be tipped in favour of the domestic animal. It may be possible to reduce the animal's susceptibility to the harmful influence of the parasite by correcting a deficient diet, stimulating immunity by vaccination, or introducing hereditary resistance factors. Manipulation of the environment can create conditions inimical to the pest, and management systems can be devised to minimize contact between livestock and host-seeking parasites or vectors.

Often direct action has to be taken against the pest population in order to cure an already sick animal or as part of an integrated control programme to prevent future losses. Biological methods such as sterile male release have been used successfully a few times, and methods for altering the genetic composition of pest populations are being tested on a small scale, but in most cases chemical control is still the only feasible method for reducing parasite numbers. In the specific case of the tick, which is among the most damaging of all veterinary pests, producing world-wide losses conservatively estimated at £200 million a year with 80 per cent of the world's cattle population at risk,[9] almost all the control measures outlined above are being applied when appropriate to local conditions. Even so, in many parts of the world satisfactory protection cannot be achieved without regularly dipping all cattle in acaricidal baths. Animals have to be treated up to ten times annually in parts of Australia and weekly dipping can be necessary in Africa.

In recent years, considerable advances have been made in the development of veterinary insecticides, acaricides, molluscicides, anthelmintics, and antiprotozoal agents, so that now at least one effective compound exists for virtually every major pest affecting animal production. But to expect a rapid eradication of all parasitic disease on this account would be erroneous, for political, economic, social, and manpower constraints, together with the continuing need for further research and education, make this a distant objective. In addition, host-parasite-environment interactions are generally complex, and control measures rarely encompass all the multiplicity of routes that the pest may use to complete its life-cycle. The emergence of strains of parasites resistant to the effects of specific chemicals also tests the ingenuity of chemists seeking new types of pesticide (see Chapter 16).

*          *          *

A successful control programme must be built on a detailed knowledge of the biology of the parasite and the reactions of the host to parasitism, together with an understanding of the local epidemiology of the disease in the social, political, and economic context of the area. Drugs, pesticides, and vaccines can then be used effectively, economically, and safely as part of an integrated scheme making use of improved husbandry and management techniques. Great efforts are being directed at international, national, and local levels towards the planning and execution of suitable control schemes, with international organizations, government bodies, industrial concerns, the veterinary profession, and many others, including the farmer himself, making real contributions. There is hope that this cooperative action will progressively diminish the enormous wastage of natural resources caused by parasitic disease.

## References

1  BEESLEY, W.N. Economics and progress of Warble Fly eradication in Britain. *Vet. Med. Rev.* **4**, 334 (1975).

2  CHERRETT, J.M., FORD, J.B., HERBERT, I.V. and PROBERT, A.J. *The control of injurious animals.* English Universities Press (1971).

3  CUMMING, J.N. An estimate of costs to the sheep industry due to blowflies and parasites. *Q. Rev. Agric. Econ.* **17**, 197 (1964).

4  FINELLE, P. African animal trypanosomiasis. Part IV, Economic problems. *Wld. Animal. Rev.* **10**, 15 (1974).

5  FROYD, G. and McWILLIAM, N. Estimate of the economic implications of fascioliasis to the United Kingdom livestock industry. *Proc. 20th Wld. Vet Congr.* 217 (1974).

6  GORDON, H. McL. Parasite penalties on production. *Proc. Aust. Soc. Anim. Prod.* **10**, 180 (1974).

7  KENDALL, S.B., Chemical treatment of coccidiosis. In *Veterinary pesticides.* S.C.I. Monograph **33**, Society of Chemical Industry, London (1969).

8  POTTER, C. The effect of pesticide usage on the quality of life. *In* Economic and social values in the assessment of crop protection and pest control methods. *Proceedings of the International Conference of the Society for Chemical Industry Pesticides Group,* 1973 (in press).

9  SHAW, R.D. Tick control on domestic animals. *Trop. Sci.* **11**, 113 (1969); 12, 29 (1970).

10  SPRINGELL, P.H. The cattle tick in relation to animal production in Australia. *Wld. Animal Rev* **10**, 19 (1974).

11  STEVENS, A.J. The thin sow syndrome. *Agriculture, Lond.* **74** (11), 510 (1967).

12  TAYLOR, E.L. Economic parasitology. *Vet. Rec.* **74** (30), 844 (1962).

# 10 The importance of pesticides in developing countries

by A. V. Adam

Were it not for impressive advances in agricultural technology, led mainly by work in the industrialized countries during the past three decades, hunger would be stalking far more people than it actually does. Even so, FAO reports that as many as 1500 million individuals still suffer some degree of hunger or malnutrition. Most of these people live in developing countries,* largely in the tropical or subtropical zone, occasionally referred to as the 'hunger belt'. They constitute about two-thirds of the human population, but produce less than one-fifth of its food. More than 50 per cent of the world's most needy people are peasant farmers, numbering about 1000 million and living in developing countries; and the peasant farmer population is expected to increase to nearly 1500 million by 1985.

Some two-thirds of the total population of developing countries depend either directly or indirectly on agriculture for survival. Furthermore, many other important sectors of such countries' economies (trade, services, and manufacturing, for example) depend to various extents on agriculture. Consequently, it is mainly through agricultural development that the growing rural population can maintain itself and at the same time help the national economy. Agriculture in developing countries is relied upon first to provide subsistence for a substantial segment of the rural and urban population and, second, to earn foreign exchange through the export of a small number of agricultural

---

*Although there are over 80 nations classified as developing countries, in 1975 FAO listed the following as the 'most seriously affected':
Afghanistan, Bangladesh, Burma, Burundi, Cape Verde Islands, Central African Republic, Chad, Dahomey, Democratic Yemen, Egypt, El Savador, Ethiopia, Ghana, Guinea, Guinea-Bissau, Guyana, Haiti, Honduras, India, Ivory Coast, Kenya, Khmer Republic, Laos, Lesotho, Madagascar, Mali, Mauritania, Mozambique, Niger, Pakistan, Rwanda, Senegal, Sierra Leone, Somalia, Sri Lanka, Sudan, Uganda, United Republic of Cameroon, United Republic of Tanzania, Upper Volta, Western Samoa, Yemen.

Plate 10.1. A swarm of desert locusts, before and after aerial application of insecticide. If not controlled, a typical large swarm can consume up to 3000 tonnes of plant material a day. (FAO photographs)

commodities ('cash crops') produced either on individual farms or under cooperative arrangements.

At the current population growth rate about 15 million additional people have to be fed annually in the Third World. Yet, thanks to the introduction of modern agricultural technology, consisting mainly of improved high-yielding varieties of cereals, countries like India have miraculously come close to the point of self-sufficiency in food grains. Such successes do not depend solely on the isolated introduction of a key input, such as an improved variety. In effect, the final outcome depends heavily on a 'package' of mutually dependent inputs, the most critical of which are seeds, irrigation, fertilizers, and plant protection (principally pesticides). It should, perhaps, be pointed out that heavy reliance on chemicals had led to the point where the word 'pesticides' has in many instances been used synonymously with 'plant protection'. One of the main reasons for this, in addition to the normally spectacular returns for the money invested, has been the relative ease with which farmers could learn to use pesticides and could use them with minimum reliance on continued guidance.

Since the publication of *Silent Spring* in 1962, the total use of pesticides world-wide has more than trebled, but in the mid 1970s all developing countries together utilized only about 8 per cent of total world pesticide production, and most of it went to public health campaigns to control disease vectors.

In the developed countries, particularly in North America, Western Europe, and Japan, pesticides have come to play an extremely important if not irreplaceable role in the maintenance of high agricultural productivity (see Chapter 5). The cost of the pesticide input in these countries is estimated at about 6 per cent of the costs of all purchased inputs. In the developing countries, where virtually all pesticides have to be imported at the expense of limited foreign exchange, the cost of the pesticide input in relation to other inputs may reach as much as 40 per cent. Accordingly, considerable care is required in selecting the compounds most suitable for local use, keeping cost benefit factors very much in mind. As a rule pesticides must be evaluated in terms of their potential usefulness either in protecting food crops or export cash crops. Countries with limited foreign exchange available for the importation of pesticides may give first priority to the protection of their export crops even at the expense of food crops. Cotton pest control provides a vivid example of large outlays on insecticides, representing approximately 24 per cent of all insecticides used throughout the world.

The relationship between the actual cost of using a pesticide and the value of increased yield and quality is a very complex one. Accurate estimates would call for much more precise information than is currently available, including the influence of each of a wide range of factors affecting yield and quality. As accurate methods for measuring loss of yield in specific crops due to a particular disease, pest, or weed have not yet been adequately developed, it is difficult to estimate what a crop might yield under disease-free and insect-free conditions, or when no control measures are applied, or to determine losses which might still occur in spite of the application of pesticides.[5]

**Examples**

The examples of losses shown in Table 10.1 can thus represent only estimates. Reliable data on crop losses are fundamentally important to governments in operating and evaluating plant protection programmes, to the pesticides industry, to international assistance agencies, and to individual farmers.

Most crop losses from pests, diseases, and weeds go unreported, but the following are a few of the better-known examples of serious losses occurring in various developing countries owing to inadequate control. Cramer has covered the subject in considerable detail.[4]

*Brazil.* Only 20 per cent of the cultivated area received pesticide treatment of any sort in 1973 and about 33 per cent of the potential crop yield was lost by insect damage, plant diseases, and weed

Table 10.1. *Estimated average losses of potential crop yield for 1973*

| Latin America | Loss % | Africa | Loss % | Asia* | Loss % |
|---|---|---|---|---|---|
| Maize | 40 | Ground nuts | 33 | Rice | 30 |
| Rice | 25 | Cotton | 23 | Cereals | 32 |
| Small grains | 21 | Cocoa | 51 | Cotton | 24 |
| Cotton | 33 | Coffee | 32 | Vegetables | 47 |
| Sugarcane | 26 | Bananas | 36 | Fruits | 39 |
| Coffee | 40 | Vegetables | 61 | | |
| Bananas | 29 | Fruits | 64 | | |
| Vegetables | 44 | Palm oil | 19 | | |
| Fruits | 42 | | | | |
| Beans | 46 | | | | |

*Excluding Japan.

118

competition. The cotton crop makes use of about 70 per cent of the total volume of pesticides employed in the country and 92 per cent of the DDT utilized for agricultural operations.

*India.* Crop losses in 1974 due to pests were valued at 50 billion rupees, according to the Indian National Council of Applied Economic Research. Rodents destroyed 6 per cent of standing crops and 6·8 per cent of stored harvests, amounting to a loss of 12.5 million tonnes of grain in 1973-4. India's total food grain import for 1974 was 4.9 million tonnes.

*Mexico.* In tropical and sub-tropical areas, weeds are known to reduce yields by over 50 per cent if they are allowed to compete for only four weeks.

*Chile.* Loss of potato yields due to late blight averaged 25 per cent in 1973-5.

Looking now at the positive side, the following examples are of interest, even though some of them are several years old.

*Pakistan, the Philippines and Brazil.* The use of selective weed killers in controlled plots has increased rice yields by 45-51 per cent compared to conventional weed control.

*Ghana.* Cocoa production was increased three-fold in 1971-3 by adequate control of capsids with insecticides.

*Sudan.* Cotton yields in 1969 were increased threefold through effective pest control.

*Uganda.* A 74 per cent increase in cotton yields was achieved in 1968 by adequate pest control.

*Pakistan.* Sugar cane production has gone up by one-third as a result of pesticides being used to prevent losses (an expenditure of about US $77 000 on insecticides made it possible in 1971 to harvest $7.2 million-worth of sugar).

*World-wide.* FAO estimates that an average of 38 per cent of the cotton crop is saved from destruction by pests through the effective use of insecticides.

## Growth in pesticide use[6,7]

Pesticide consumption for 1973, reported by 38 developing countries for which data were sufficiently detailed to allow analysis, totalled 160 000 tonnes in terms of active ingredients: 106 000 tonnes of insecticides, 49 000 tonnes of fungicides, and 5000 tonnes of herbicides. In 1975 the same 38 countries reported a total consumption of 202 000 tonnes.

Increases in the rate of consumption of herbicides exceeded those for all other classes of pesticides. This was due in part to the recent introduction of high-yielding varieties of rice and wheat, which are more reliant on fertilizer and weed control than are traditional varieties. Recent data received from some of the developing countries indicate that yields from fertilized, high-yielding wheat varieties were depressed from 30 to 80 per cent when the crops were not treated with herbicides. Similar data have been received for rain-fed rice in Africa and Asia.

Mainly as a result of higher prices, the rate of increase projected for the years 1975-7 is considerably less than for the previous three years, representing only a 9 per cent compound increase per annum. In this period herbicides will account for some 10 per cent of total pesticide use in developing countries. Although they represent only a minor part of total pesticide use, the importance of herbicides in the successful production of high-yielding grains makes their supply important.

In the Central American countries, where pesticides have been available for a relatively long period, the overall rate of increase in pesticide use (17 per cent per annum) was lower than in South America (22 per cent) or Asia (35 per cent). As pesticides have been a relatively newer input in the Asian countries, it can be expected that their future rates of growth will continue to exceed those of South and Central America. The African countries surveyed showed a net decline in pesticide use during the 1971-3 period, brought about by an annual decrease of 11 per cent in insecticide use, which was probably due to a lack of available supplies and foreign exchange rather than to a decrease in demand.

### Aspects of pesticide use

A series of regional seminars held between 1971 and 1975 for national heads of plant protection services, public health authorities, technical representatives from major pesticide manufacturers, and FAO and WHO personnel in Latin American, Asian, and African countries reached a remarkable consensus of opinion on basic issues involving the current and future use of pesticides in developing countries.[1,2, 3, 11] The principal conclusions reached at these meetings are of considerable importance and it is worth summarizing the main points here.

### *Importance of food losses*

As mentioned above, about two-thirds of the world's population live in developing countries, where the rate of increase in food production lags substantially behind the increasing population growth, and where very serious losses in food production are occurring in spite of some use of pesticides. High-yielding varieties have generally proved more susceptible to attack by pests and diseases and require greater use of fertilizer, which, in turn, tends to enhance their susceptibility to plant diseases and to promote greater growth of weeds. It is generally considered that average world-wide pre-harvest food losses in developing countries are about 30-50 per cent of the harvested crop. An additional 5-15 per cent loss is attributable to post-harvest problems. Furthermore, animal production, which is extremely important to many developing countries, is severely hampered by external and internal parasites that are known to reduce output severely. The use of pesticides on domestic livestock and for the control of vectors is therefore important (see Chapter 9).

*Basic considerations*

About 69 per cent of the total work-force in developing countries is employed in agriculture, including forestry, and agriculture provides on average some 52 per cent of the national income and 73 per cent of the foreign exchange earnings, excluding the petroleum and mineral exports of these countries. Yields on farms are greatly below their potential, one of the principal causes being the ineffective control of pests and diseases, and there is an urgent need for the use of pesticides and other plant protection measures to be substantially increased. The

Table 10.2 *Variation in use of plant protection chemicals**

|  | Active ingredient applied (g/ha) |
| --- | --- |
| Japan | 1233 |
| Korea | 214 |
| Sri Lanka | 145 |
| Mexico | 55 |
| India | 50 |
| Brazil | 29 |
| Philippines | 27 |
| Egypt | 24 |
| Kenya | 11 |

*Figures are for 1973 and are based on total quantities of pesticides used vs. total area treated.

huge variation in the average quantities of plant protection chemicals, in terms of grams of active ingredient applied per hectare of cultivated land, among certain countries is of interest, and has been calculated as shown in Table 10.2.

*Safety*

Priorities in solving problems associated with the use of pesticides differ greatly between developed and developing countries, the latter being mainly interested in solving the problem of intoxication as a result of direct contact, whereas the former are mainly interested in residues and environmental issues. Currently (1976), one of the important problems connected with the use of pesticides in most developing countries is the incidence of intoxication and death, resulting mainly from the misuse of certain products. No reliable statistics exist on the incidence of pesticide poisoning or on medical services in rural areas. For various reasons, affected people often fail to report illness, and deaths may also go unreported. From the known cases, however, acute poisoning from pesticides, due either to mishandling of the chemicals or to accident or to suicide, is evidently significant in most developing countries.

Problems associated with pesticide containers are of considerable concern in developing countries. They include unsafe packing and the hazard of leakage in storage and transport, with consequent risks of contamination of food and clothing; the strongest possible measures are needed to prevent such accidents. The effect of heat and moisture on many packaging materials also requires careful consideration; and the disposal of used containers presents special problems in developing countries where people may save or sell empty pesticide containers for transporting, storing, or preparing food and drink. In communities where people are largely unfamiliar with acute poisons intoxication often results from highly toxic substances being kept in inadequately labelled drinking bottles and food containers.

In a few of the less developed countries too many very toxic pesticides are available. A few carefully selected products could be adequate and present a reduced hazard. Agricultural authorities ought, as far as they can, to inform health authorities of the range of pesticides being used, so that doctors can familiarize themselves with symptoms of intoxication and with the best treatment. Existing public health facilities should be used, and they should be equipped to deal with cases of exposure to toxic substances. Where the use of pesticides is most intense, as in cotton production areas, principal users should where possible be monitored for evidence of excessive exposure. Having said all this, we must recognize that the hazard from pesticides arises more from ignorance, carelessness, and negligence than from the inherently toxic nature of many widely used materials.

## Formulation

Formulation is very important in ensuring the effective and safe use of pesticides. Tropical conditions pose particularly difficult problems involving the stability, physical condition, and otherwise satisfactory quality of products. Although most major manufacturers devote considerable research to providing adequate formulations, there have been some costly (and occasionally injurious) failures. It can sometimes happen that local formulators do not maintain adequate standards; this may be partly due to inexperience and to inadequate facilities and technical knowledge, but there are occasional cases of adulteration and misrepresentation. The need for proper standards for the chemical, physical, and biological properties of products on local markets is of great importance.

Most developing countries are anxious to encourage the establishment of local formulation facilities; for this, technical assistance is required with equipment, expert services, local availability of raw materials, quality control, protection of operators, and prevention of pollution. And this cannot be achieved without the cooperation of basic chemical manufacturers, governments, and international technical agencies.

## Pesticide application

The effective use of pesticides depends to a great extent on how they are applied. Much of the equipment used in developing countries is poor, and often there is not enough of it. Some makes and designs have proved unsuitable and some require excessive maintenance which often cannot be provided locally. The quality of the equipment is certainly important, but simplicity of design is also needed: simple, hand-operated knapsack sprayers requiring minimal maintenance are the only machines that have proved useful for wide-scale use on farms in developing countries; but it is, of course, important that spare parts should be available at reasonable prices.

If practical aspects of pesticide application, such as selection of equipment, calibration, particle size, and field mixing, are not properly understood, or dealt with, the results may include ineffective control of pests and diseases; extremely high application costs; waste of expensive and potentially danagerous chemicals; objectionable pesticide residues; possible contamination of the environment; damage to crops; and intoxication hazards.

One should add that failure to control a certain pest or disease is often quickly, but erroneously, attributed by farmers to the ineffectiveness of a pesticide. In an attempt to correct the error, farmers frequently adopt changes in their pest control practices (including 'over-dosing') which aggravate rather than alleviate the original error.

## Persistent pesticides

A dilemma is being faced by several developing countries over the use of DDT and certain other persistent organochlorine compounds. In some countries such compounds are being banned because of action in other regions and because of adverse publicity. While recognizing that persistent pesticides should wherever possible eventually be replaced by less persistent materials, the decision should be based on technical grounds, with full recognition of the consequences of introducing substitutes which can be more hazardous if used by uninformed farmers and which are, as a rule, more costly. Furthermore, the persistence of residues in the environment is of distinctly shorter duration in tropical than in temperate areas, a fact which tends to reduce the importance of this property.

## Residues

Residues of pesticides in food, raw agricultural commodities, people, and components of the environment have not so far been studied extensively in most developing countries, for the necessary facilities and trained personnel are not yet available, nor have any hazards appeared. In view of the cost of equipment, the many technical difficulties, and the shortage of experienced residue chemists, there would be advantages in developing laboratory facilities for residue analysis on a regional basis.

Occasional difficulties in trade in some commodities with certain countries, resulting from the use of particular pesticides or the presence of unacceptable residues in the importing countries, have accentuated the importance of having proper facilities for residue analysis in the exporting countries. Although most countries are still at an early stage in the development of residue monitoring, FAO strongly recommends that as many individual countries as possible should provide information on all relevant data to help the Joint FAO-WHO Meeting of Experts on Pesticide Residues to make recommendations which take account of the special requirements of the countries concerned and reflect the needs of producers and exporters.

Although some users do not understand or follow good agricultural practices in the use of pesticides, and monitoring is required, it is generally felt that legislation and quality control facilities to regulate the import, sale, labelling, and quality of pesticide formulations should take priority over monitoring. World-wide experience during the last 30 years has shown no evidence that consumers have been injured by residues resulting from approved application of pesticides, even in regions where the rate of application was many times higher than could be expected for most developing countries.

## Environment

Professor Mellanby discusses questions of pesticides and the environment in some detail in Chapter 18, but the subject merits brief reference here in connection with the developing countries. Environmental pollution generally, and pollution from pesticides in particular, have received a good deal of attention, but although some developing countries have appointed commissions to advise on and regulate pollution, pollution from pesticides has not generally been considered as a priority under present and foreseeable circumstances. In fact, concern has been expressed by a number of developing countries that environmental problems, not yet of crucial importance in most of them, might be given such high priority as to overshadow the critical need for agricultural development. This, however, does not detract from general concern about possible future adverse effects, and the need for vigilance and continuous review of new technical information and experience, especially from developed countries.

## Legislation

Basic pesticide legislation and official control procedures at the national level are essential first steps in promoting the safe and effective use of pesticides. The need for uniformity of pesticide legislation and of approaches to problems associated with the registration of pesticides, and the establishment of licensing schemes has been stressed, by both government officials and industry (see Chapter 19).

There has been a proliferation of uncoordinated national restrictions on pesticide registration requirements and hence availability and use; this has led to requests for regional or even world-wide action to bring about the greater uniformity which is essential to reducing the cost of research and development and thus the cost of pesticides. By the time this book is published, FAO will probably have convened an inter-governmental meeting to standardize, where possible, registration requirements internationally.

Comprehensive legislation to control importation, sale, labelling, packaging, and, in some instances, availability of pesticides, vital to ensure both effective use and safety is very important in developing countries, 40 per cent of which are estimated to have no specific legislation: even where they do, individual schemes are inadequate.

Unsatisfactory and misleading labelling of some pesticides, especially those offered by local formulators and re-packers, is a problem in many developing countries. While most major manufacturers label their products carefully, deficiencies in the precautions, limitations, and directions have been known. More detailed and more specific directions and training, and strong laws to control labelling, are thus required. To ensure the required quality and stability, realistic specifications should be adopted and purchasing agencies should buy only products meeting specifications such as those published by FAO for agricultural uses and by WHO for public health uses. Regulatory authorities should legislate for similar specifications to protect not only users but also competitive manufacturers who are prepared to produce satisfactory, stable formulations. Better storage should be developed wherever possible, but products that cannot retain full potency and physical condition for a reasonable period, at temperatures normally encountered in the countries concerned, should be prohibited.

Official control should be vested in ministries of agriculture, but in order to integrate the legislation with extension and development services it is essential that there should be close cooperation between ministries of agriculture, health, commerce, and industry, and trade at local, national, and regional levels.

*Integrated control*

The desirability and actual need to adopt an 'integrated pest management' approach to solve certain problems inherent in using pesticides is evident. It is obviously important to stabilize the pest population and to favour the development of natural mortality factors including parasites, predators, and pathogens and to adopt cultural practices against pests and diseases. Such measures would prolong the useful life of valuable pesticides, slow down the development of resistant strains, reduce residues, prevent the development of secondary pests, and possibly reduce costs in the long run.

In order to undertake integrated pest management it is necessary to have a great deal of fundamental biological data, selective pesticides, adequately trained specialists and technicians, farmers who are receptive to new concepts, and ample statistics on crop losses and economic threshold levels for each crop. Many of these requirements cannot easily be met in highly developed countries, let alone in developing countries; all the same, the basic principles of integrated pest management should be applied wherever and whenever possible. Having said this, we must remember that the basic, all-embracing objective in developing countries is to restrict losses as effectively as possible through the use of the best practical means available at the time. Unfortunately some developing countries have been led to believe that pest management principles can be readily adapted to their own conditions long before they are in a position to adopt them.

**Training and extension**

Training in plant protection and particularly in the proper use of pesticides is a high priority in all developing countries. In the past, extension efforts have frequently been handicapped by the lack of basic information, shortage of trained technicians, need for motivation, lack of technical direction, and low literacy. Communication, including the ability to train, is influenced to a great extent by the level of literacy, and so training farmers calls for an effort to reduce or eliminate illiteracy among them.

The critical need for effective extension work is amply demonstrated when one realizes that less than 5 per cent of the available technical information on plant protection reaches and is actually used by farmers in developing countries. To improve this alarming state of affairs, manufacturers, distributors, and government should cooperate to train the user in proper pesticide handling and application techniques. They could help by placing on the local market only chemicals which are appropriate to the country, providing easy-to-understand literature, setting up demonstration plots, sponsoring training lectures to groups of farmers, assisting growers with problems relating to application equipment, and assuring delivery of suitable formulations at the right time. To be able to do these things, the private sector needs the active support of governments, and vice versa.

There is no question that an understanding of the psychological barriers to the acceptance of new ideas by members of a rural community has frequently determined the final success of a training programme. Farmers and their families are not necessarily impressed by communication techniques which have proved successful in towns. The involvement of local people and experience and the use of familiar traditions and practices can help farmers to identify themselves more easily and rapidly with the message that the trainer is trying to convey.

### Constraints

Snelson[12] has outlined the following 27 potential constraints which can inhibit the introduction, distribution, and optimum use of pesticides:

    inadequate socio-economic structures;
    factors that impede agricultural development;
    lack of essential inputs for agriculture;
    lack of rural infrastructure;
    inadequate supporting services, education, and extension;
    unavailability of credit and incentives;
    shortage of dedicated leadership;
    trade restrictions;
    inadequate plant protection services;
    lack of knowledge of pests, diseases, and weeds;
    shortage of technical and managerial 'know-how';
    lack of local applied research;
    illiteracy, suspicion, tradition, and communication problems;
    inadequate wholesale or retail distribution;
    unrealistic local registration requirements;
    shortage of foreign exchange;
    fears of consequences of environmental contamination;
    supposed dangers of residues in food;
    low maximum residue limits;
    absence of reliable pest and crop-loss information;

ignorance;
lack of reliable guidance on safe practices;
development of resistance by pests;
fear of pesticide poisoning;
emotional publicity against pesticides;
demands for more and more sophisticated studies;
price.
This comprehensive list could prove useful as an aid in planning and
evaluating pesticide programmes, especially in developing countries.

*        *        *

Pesticides, particularly when accompanied by improved seeds,
fertilizers, and irrigation facilities, are known to contribute significantly
in reducing losses and thereby increasing food production and quality.
Past trends indicate, however, that most of the pesticides used in
developing countries have frequently been employed in the control of
vectors of human disease and on industrial export crops, whereas most
of the uses in developed countries are on food crops: it is in this sphere
that pesticides must be used increasingly in the developing countries.

Improved crop protection, accomplished mainly through the effective
use of pesticides, not only increases crop yields but also leads to better
health and a higher standard of living. Accordingly, large increases in
the use of agricultural pesticides will be necessary in developing
countries if they are substantially to increase food production. In fact,
it has been estimated that a fivefold increase by 1985, compared with
1965 use levels, will be required before a real impact is felt in food
production and availability through a significant reduction in pre-
harvest and post-harvest food losses.

Priorities and objectives in the search for new pesticides are changing,
and it is logical to assume that new products will meet more closely the
increasingly strict criteria that relate to safety and the environment and
will possess properties conducive to their use under pest management
systems.

Past trends and experiences on an international scale in developing
countries have shown beyond doubt that the judicious use of agricultural
pesticides has made a major contribution to the production of food and
fibre, and has thus been instrumental in increasing economic development and stability and in improving personal and national freedom. We
can expect that it will continue to do so for many years to come and on
an expanding scale.

## Further reading

1   ADAM, A.V. *Report of FAO/Industry Seminar on the safe and effective use of agricultural pesticides in South America.* FAO: DDI:G/71/17 (1971).
2   —— *Summary of Joint FAO/Industry Seminar on the safe, effective and efficient use of pesticides in agriculture and public health in Central America and the Caribbean.* FAO:ACPP:MISC/6 (1972).
3   —— *Report of FAO/ICP Seminar on the safe, effective and efficient use of pesticides in agriculture and public health in Asia and the Far East.* FAO: AGPP:MISC/74/12 (1973).
4   CRAMER, H. Plant protection and world plant production. *Pfl Schutz-Nachr. Bayer,* **20,** 1-524 (1967).
5   *FAO Manual on the evaluation and prevention of losses by pests, diseases and weeds.* Supplement 1. Commonwealth Agricultural Bureaux, Farnham Royal (1971).
6   FAO. Various internal reports and documents. 1950-73.
7   —— Ad hoc *government consultation on pesticides in agriculture and public health.* Background and working documents, FAO:AGP: 1975/M/3, 1975.
8   —— *Provisional indicative world plan for agricultural development.* Rome (1970).
9   —— *The state of food and agriculture.* Rome (1970).
10  *Pest control strategies for the future.* National Academy of Sciences, Washington, D.C. (1972).
11  SNELSON, J.T. and ADAM, A.V. *Report of FAO/ICP seminar on the efficient and safe use of pesticides in agriculture and public health in Africa.* FAO:AGPP:MISC/18 (1974).
12  —— Prospects for use of existing and new pesticides: Major factors limiting introduction, distribution and optimum use. *FAO Backgr. Pap.* AGP: PEST/PH/75/B48 (1975).

# 11 Rice

by M. D. Pathak, S.H. Ou, and S. K. De Datta

Rice is the world's most important food crop and is the staple diet of over half the world's population. It is grown on over 130 million hectares, the largest area under any single crop. Ninety per cent of the total area under rice is in the developing tropical and sub-tropical countries, where the population densities are among the highest in the world and where incomes *per capita* are among the lowest.

Classified primarily as a tropical and sub-tropical crop, rice is cultivated as far north as Czechoslovakia and Hokkaido, Japan, and as far south as the North Island of New Zealand and Paraguay in South America. It is grown from sea level to altitudes of about 3000 metres in the Himalayan mountains. Its optimum temperature requirements range between $20^{\circ}$ and $30^{\circ}$C. It is a water-loving plant — if the total water needed by a rice crop of 130-50 days duration were placed all at one time in the field it would be about a metre deep.

Throughout the approximately 5000 years of domestication, the rice plant and the systems used to cultivate it have been adapted to existing local conditions. In the plains it grows in bunded fields which impound the rain or irrigation water. On rolling lands, it grows as a rainfed upland crop without standing water, and on mountain slopes it is grown in picturesque terraces. It is produced in flood plains where waters may be as deep as 5 m after the rains in the adjacent highlands. No other staple food crop under cultivation has such a wide range of adaptation.

In the wet tropics, rice is known for its ability to grow under conditions where no other crop will grow. However, the yield of native varieties of tropical rice is low, because they are excessively tall and therefore lodge easily. Thus, their potential for effectively utilizing natural soil fertility or added fertilizers is restricted. In addition, native varieties of rice are subject to frequent epidemics of pests and diseases and their optimum performance depends on high rainfall or on irrigation. As a consequence, rice yields in the tropics and sub-tropics range from 1.5 to 2.0 tonnes per hectare, compared to yields of up to 6 t/ha that are common in temperate zones where conditions are favourable for

Plate 11.1. As is shown in Fig. 11.1, insects and diseases are a major limiting factor in rice production. Here a Javanese farmer sprays his rice seed bed against a rice borer; stem borers are among the most common insect pests of the rice crop. (FAO photograph by F. Botts)

controlling pests, for water and nutrient supplies, and for the use of improved cultural practices.

### Recent improvements in rice production

In the early 1960s, scientists began to make progress in genetically restructuring the tropical rice plant to overcome some of these weaknesses. Their aim was to develop a plant that was short and stiff-strawed, heavy tillering with moderately upright leaves, and not sensitive to the changes in day length. This plant type was to be bred to use sunlight, water, and fertilizer nutrients efficiently.

One of the products of these efforts was IR8, a variety developed in 1966. (The prefix IR is used for varieties developed at the International Rice Research Institute). The performance of IR8 was outstanding: 6 to 10 t/ha yield as compared with 3 to 5 t/ha with traditional varieties in many areas. It became known as 'Miracle Rice' and has been referred to as an initiator of the Green Revolution in rice. However, in areas such as Bangladesh, Sri Lanka, some parts of India, Indonesia, and the Philippines, where insect and disease problems were severe, the success of IR8 was limited.

Realizing the severity of insects, diseases, and environmental factors, specialists of different scientific disciplines worked in teams to incorporate into high-yielding rice varieties genetic resistance to as many of these hazards as possible, particularly insects, diseases, drought, problem soils, floods and deep water, and abnormal temperatures. The results of such teamwork were encouraging (see Table 11.1) and the future prospects are even brighter.

The rice varieties developed from such interdisciplinary team efforts provide improved yield stability even under adverse agro-climatic and pest conditions, thereby reducing farmers' risks. This encourages farmers to adopt these new rices and to invest in the inputs to realize their full yield potentials. The success and adoption of some of these varieties are evidenced by IR20, which within three years of its release was planted on five million hectares throughout south-east Asia. While vast areas planted to susceptible varieties were devastated by the tungro virus, the adjoining IR20 fields were lush and green and produced high yields. Similarly, IR26, which is resistant to the brown planthopper *(Nilaparvata lugens),* suffered little harm by this insect in some parts of the Philippines and Indonesia, while susceptible varieties were severely damaged.

Although crop resistance to insects and diseases provides, perhaps,

Table 11.1 Resistance ratings of IRRI varieties*

| Variety | Diseases | | | | Insects | | | | Soil problems | | | |
|---|---|---|---|---|---|---|---|---|---|---|---|---|
| | Blast | Bacterial blight | Grassy stunt | Tungro | Green leaf-hopper | Brown plant-hopper | Stem borer | Gall midge** | Alkali injury | Salt injury | Zinc defi-ciency | Phos-phorus defi-ciency |
| IR8 | MR | S | S | S | R | S | MS | S | S | MR | S | MR |
| IR5 | S | S | S | S | R | S | S | S | S | MR | R | MR |
| IR20 | MR | R | S | R | R | S | MR | S | S | MR | R | R |
| IR22 | S | R | S | S | S | S | S | S | S | S | S | MR |
| IR24 | S | S | S | MR | R | S | S | S | MR | MR | S | MR |
| IR26 | MR | R | MS | R | R | R | MR | S | MR | MR | S | R |
| IR28 | R | R | R | R | R | R | MR | S | MR | MR | R | R |
| IR29 | R | R | R | R | R | R | MR | S | S | MS | R | R |
| IR30 | MS | R | R | R | R | R | MR | S | MR | MR | R | MR |
| IR32 | MR | R | R | R | R | R | MR | R | S | – | – | – |
| IR34 | R | R | R | R | R | R | MR | S | S | S | R | R |

R, resistant; MR, moderately resistant; MS, moderately susceptible; S, susceptable. * Rated in the Philippines. ** Rated in India.

the most significant means of controlling rice pests, insect and disease problems are so numerous and varied that such resistance must be complemented by chemical and cultural means of control. As with other crops, integrated methods of pest management for rice offer the best long-term solutions.

## Pest problems

Next to the vagaries of weather, damage from insect pests and diseases provide the most significant constraints to rice production, particularly in the tropics. Weeds are a universal problem whether rice is grown as an upland crop, in the paddy field, or as a floating rice under deep water. Insect pests and diseases are most troublesome in humid, tropical environments, especially where climatic conditions permit the growing of rice throughout the year. Likewise, some improved agronomic practices which induce dense crop stands and succulent plant tissues sometimes favour the growth and development of insect pests and diseases.

Quantitative figures on the extent of crop damage from rice pests are not widely available. In specific cases, however, damage has been economically significant. In the Philippines, for example, the tungro virus virtually destroyed 70 000 ha in 1971, and 40 000 ha in 1972. In addition, the brown planthopper infested at least 80 000 ha in 1972. Had these epidemics not occurred, the Phillippines could easily have maintained the self-sufficiency in rice which was attained in 1969. Instead, it has had to import annually 46 000 tonnes of this indispensable food crop.

Other examples of serious pest epidemics are the tungro virus infestations in Bangladesh, India, and Thailand, and the brown plant-hopper outbreaks in India, Indonesia, and Sri Lanka. Less devastating pests, such as rice stem borers and whorl maggots, and several diseases, together with the ever-present competition from weeds, reduce yields and drastically impair national levels of rice production.

A recent series of experiments conducted by the International Rice Research Institute in farmers' fields in Laguna province, Philippines, showed inadequate disease and insect control to be the most significant constraint on farmers' yields (see Fig. 11.1). During the wet season, the main crop season, 88 per cent of the yield increase effected by improved cultural practices was attributable to the control of insects, diseases, and weeds.

Such statistics emphasize the critical role of proper pest and disease

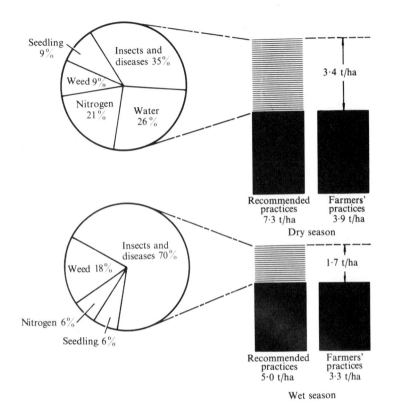

Fig. 11.1    Differences between yields when farmers follow their usual practices and when they follow IRRI recommendations. Factors which constrain yields in farmers' fields are shown in the circles. Data taken on a total of 15 farms over three crop seasons, 1972-3, Laguna province, Philippines. Source: *IRRI Research Highlights for 1973.*

control in increasing the world's rice production. A brief discussion on each of these problems follows.

*Insects*

In the tropics and the sub-tropics of Asia, more than 100 species of insects have been recorded as rice pests. Fifteen of these insects generally occur regularly and infest the crop at various stages of plant growth. Several other insect species occur only sporadically, but cause significant losses.

The insect pest problem is not severe enough in most Latin-

American countries and in Australia, where rice is a comparatively recently cultivated crop, to warrant general widespread control measures. Nor are insect pests very serious in Mediterranean countries. Although two stem-borer species (one lepidopterous and one dipterous) cause sporadic damage to the rice crop in Africa, at present most farmers employ no control measures. The rice crop in Iran was until recently relatively free of insect problems. A stem-borer *(Chilo suppressalis)* appeared in a few rice fields along the Caspian sea about 1971; by 1974 it had spread to nearly 50 000 ha. In infested areas, a good rice crop cannot be grown without pesticidal protection.

Insect problems are quite severe in Asia, which has 90 per cent of the world's rice area. The effects of insect pest damage, in addition to those of unfavourable weather conditions, on rice production during a 90-year period in Japan are shown in Fig. 11.2. Although comparable data are not available for other Asian countries, greater yield depressions

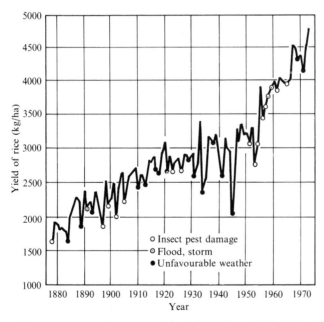

Fig. 11.2  Fluctuations in average yields of rice in Japan 1878–1974. The role of insect pests in the overall production of rice is apparent. Today virtually all farmers in Japan use pesticides in their rice fields. Source: Yoshimeki, M., in *Proc. Symp. major insect pests of the rice plant*, John Hopkins Press (1967), and Agriculture and Forestry Statistics, Tokyo (1974).

137

from pest infestation are suspected for most of the tropical and sub-tropical countries.

Evidence of the intensity of insect and disease damage can be found in data from 50 different experiments conducted during 14 years at the International Rice Research Institute located in the Philippines. Plots protected by pesticides produced an average grain yield of 5.7 t/ha while the average yield of comparable unprotected plots was 2.9 t/ha. Another set of 130 experiments conducted in farmers' fields at different locations in the Philippines during 1966-7 showed that plots protected from insect damage produced, on the average, 1 tonne more rice per hectare than the plots which received similar treatments but were not protected against insects. This amounted to about 25 per cent more rice when pest control was practised. In recent years, careful experiments conducted in farmers' fields in various Asian countries (India, Indonesia, Thailand, Sri Lanka, and others) have generally shown similar results.

Stem-borers, the most common insect pests of rice, occur throughout the world. Several species of leafhoppers and planthoppers are serious pests in many areas; besides damaging the crop by direct feeding, they also transmit virus diseases. Tungro virus transmitted by green leafhopper and hopperburn caused by the brown planthopper have damaged large areas of rice in several countries in recent years.

Rice gall midge has been another severe pest in India, Indonesia, Sri Lanka, and Thailand. This insect infests rice during its vegetative stage of growth and the growing parts of the plants are transformed into onion-leaf-like structures. The affected tillers produce no panicles, and thus no grain. Until a few years ago, there was no efficient method of controlling rice gall midge and a 30-50 per cent crop loss due to this insect was common in its indigenous areas of occurrence. Now there are efficient insecticidal control methods, and rice varieties possessing natural resistance to gall midge have also been developed.

Several species of rice bugs or stinkbugs infest rice fields when the grain is at the 'milk stage' and suck the contents. A 10-15 per cent yield loss caused by these insects has commonly occurred in many areas; under large infestations, even 80-90 per cent losses are not uncommon. Similarly, in many areas armyworms, cutworms, leaf folders, etc., although they occur only sporadically, cause severe crop losses.

The control of many of these pests is generally simple and is usually achieved with a single insecticidal spray at an appropriate time. However, many farmers in the developing countries keep losing a significant part of their crop to these insects because they lack information and capital or because chemicals are not available. This is in contrast to the situation in Australia, Japan, Spain, and the U.S.A., where insects and diseases are relatively less hazardous and where rice farmers are better informed and equipped to protect their crop where pests are a problem. Adequate insect control is essential for a sustained increase in rice production.

## Diseases

Several diseases caused by fungi, bacteria, and viruses are serious problems in rice production. Rice blast *(Pyricularia oryzae)* and sheath blight *(Thanatephorus cucumeris)* are common to all rice-growing areas. Bacterial leaf blight *(Xanthomonas oryzae)* and leaf streak *(X. trans-lucens* f. sp *oryzicola)* and tungro and grassy stunt viruses are prevalent in Asia. Stripe virus and dwarf virus are important in temperate Asia, and Hoja blanca occurs in Latin America.

Exact figures of the damage caused by these diseases are difficult to find, particularly in tropical Asia where most rice is grown. The yield losses in Japan due to blast alone were estimated to range from 1.4 to 7.3 per cent, with an average of about 3 per cent during 1953-60, in spite of the extensive use of chemicals and of resistant rice varieties to control the disease. The tungro virus caused an estimated loss of 1.4 million tonnes of rice or 30 per cent of total rice production in 1944 in the Philippines. (The disease was then called by other names.) It infected 660 000 hectares of rice in 1966 in Thailand and was the major disease of rice in Indonesia in the 1930s. Considering all the diseases affecting rice, a conservative estimate of loss in total rice production is 10-15 per cent. Modern rice cultural practices, including greater use of fertilizers and double cropping (growing two crops of rice in succession), tend to increase the incidence of diseases.

The use of chemicals to protect rice from diseases is most common in Japan. This is because there are economical incentives from the high support price for rice, and strong technical resources in forecasting and extension services, which are given both by the government and by the chemical industry. In the tropics, however, where these incentives are lacking, chemicals are used only to a very limited extent, despite the attempts of government institutions to introduce and encourage their use.

Among the diseases, blast and sheath blight can be controlled most effectively by chemicals, but no chemical has been found effective against the bacterial leaf blight. In general, virus diseases have not been effectively controlled through the elimination of vector insects, especially when the vector population is high, as the virus is transmitted before the insects are killed by chemicals. The rice crop can be protected from tungro virus infection, however, through appropriate timing and use of insecticides to control the carrier, the green leafhopper. A standard recommended set of practices has been developed and these are being followed in the Philippines and in a few other Asian countries.

Rice blast epidemics are brought about by high populations of air-borne spores of the fungus. In temperate regions, high spore population occurs for only one or two months, while in the tropics the high population occurs for six to seven months, or even during the entire year. Complete protection with chemicals during all stages of rice growth is not practicable in the tropics.

Chemicals may be used to protect the crop from the severe losses due to panicle infection referred to as 'neck blast'. Infected fields sprayed twice at weekly intervals starting just before panicle emergence may save about 1 tonne of rice per hectare in the Philippines. Sheath blight may be controlled by spraying at the same time. Some new chemicals are being found to be effective for both diseases.

The extensive use of mercury compounds for blast control in the 1950s and 1960s in Japan caused serious environmental pollution problems, but alternatives to these chemicals have been found. With primary emphasis being placed on host resistance, chemicals are likely to continue to play a supplementary role in rice disease control and management in the tropics.

### Weeds

Without weed control, high yields of rice cannot be achieved under any natural field conditions. Proper land preparation and flooding of fields are important deterrents to weed growth. Many scientists believe that the 'paddy' system of transplanting rice on puddled soils and keeping the fields flooded was evolved primarily to solve the weed problem. Cultivation to puddle the soils destroys many weeds, and many weeds fail to emerge through the standing water after trans-planting: it is popularly said that water is the best herbicide.

Weed infestation is generally most serious in fields without standing water, such as in upland rice. In some upland fields, the weed problem becomes so severe that it virtually precludes the growth of rice. Thus, weed control is probably the most serious problem of upland rice farmers. This is also true for the slash-and-burn type of culture common in Latin America and in certain parts of Africa.

Although the yields of all rice varieties are reduced by weed competition, weed control is particularly important when growing modern, high-yielding semi-dwarf varieties which, unlike the traditional tall leafy varieties, generally do not shade out the weeds. Also, weeds grow more profusely under the conditions of high soil fertility that are necessary to realize the maximum performance of these new varieties of rice.

Most farmers understand that the extent of yield losses caused by weeds depends to a major extent on the density of weeds per unit area, and they therefore practise manual or mechanical weeding. What many farmers do not understand is that even a few weeds per unit area cause substantial yield losses. Most fields tested in a 1972-3 study were well weeded according to farmers' standards (Fig. 11.1). The average yields of 3.3 and 3.9 t/ha of the farmer-managed plots of the wet-season and dry-season crops, respectively, are considered good yields. Nevertheless, control of the few weeds that were left in these plots increased yields by 0.2 to 0.3 t/ha. Ten per cent is probably the minimum yield loss caused by weeds even in fields that many farmers consider 'well-managed' in tropical and sub-tropical countries.

A 1973 study conducted by the Joint Commission on Rural Reconstruction, Taiwan, revealed that in an accelerated rural development programme in which 14 986 ha of rice were treated with herbicides 12 258 ha showed yield increases equivalent in value to an average of $ 10/ha over that of hand weeding.

Direct-seeded rice, if unweeded, is easily overtaken by weeds and losses are high. Losses are also substantial even in a transplanted crop of rice when weed control operations are neglected after transplanting.

No quantitative data are available on yield losses due to weeds on zonal or national bases. Yield losses from weeds have, however, been reported to range from 10 to 50 per cent in experiments on transplanted rice and from 50 to 90 per cent on upland fields. In general, yield losses due to weeds are highest in the developing countries where effective weed control is not a standard agricultural practice.

Proper land preparation and, wherever appropriate, maintenance of standing water in the fields are the most basic method of weed control. These are supplemented by hand weeding, and the use of manual or machine-operated weeders and herbicides. The appropriate use of certain herbicides provides effective weed control at a cost even lower than that of hand weeding.

While most weed species common in rice fields are easy to control, those weeds which propagate by underground tubers and rhizomes are not. For some weeds, effective chemicals are not known and hand weeding or the use of weeders which frequently subdivide the rhizomes often further compound the problem. At present, two such weeds – *Scirpus maritimus* in lowland rice and *Cyperus rotundus* in upland rice – are serious problems in certain areas in Asia.

The grassy types of weeds are also difficult to control because most grass-killing herbicides are also toxic to rice, which is also a grass. In addition, young seedlings of certain grassy weeds closely resemble rice seedlings, making hand weeding a problem. At present large areas of rice lands in Brazil are no longer planted because of the serious infestations of the red rice weed.

Effective weed control is part of good agronomic practices, both cultural and chemical. In developed countries where labour is expensive, herbicides are used. In Korea and Taiwan, only 10 per cent of the fields were treated with herbicides just a few years ago, but herbicides are now being used on about 50 per cent of the fields. This change has occurred because of the increasing cost of the labour. A similar trend is also developing in the tropics, where farmers with more than 2-3 ha, who hire labour for hand weeding, are beginning to use herbicides. Of course, farmers with even larger holdings find hand weeding much more cumbersome than the use of herbicides.

The use of herbicides is much more practical than hand weeding on upland rice. Hand weeding is very time-consuming, and since upland rice is mostly broadcast seeded and there are no rows, mechanical or manual weeders cannot be used. On lowland rice, which is planted in rows, manual weeding and herbicides are, however, equally effective.

*     *     *

The rice crop is subject to a wide variety of insect pests, diseases, and weed problems. In most of the major rice-growing areas, acceptable rice yields cannot be obtained without pest control. The use of chemicals to control these pests is one of the major practical approaches. Generally, it is not a question of justifying the use of chemicals but of determining how this need can be met. Affluent farmers employ pesticides and harvest better yields as a result. Poor farmers find it more difficult to obtain and apply these chemicals.

The scientist's approach is to develop varieties with greater yield stability and to utilize control methods which reduce the need of chemicals. They feel that more fields would be protected if the number of treatments needed could be reduced to only one or two. Even then the quantity of pesticides needed would be immense, with a potential of 80-90 million hectares needing that protection.

Plate 12.1. Symptoms of blister blight on a leaf of the tea plant. This disease, which is caused by the fungus *Exobasidium vexans*, first appeared in India and Sri Lanka (then Ceylon) in 1946. Spraying against blister blight with copper fungicides can save a tea crop from total destruction. (ICI photograph)

Plate 12.2. Picking coffee beans in Kenya. In recent years probably the most destructive disease of the coffee crop in East Africa has been coffee berry disease (CBD). Chemical control of the disease can prevent enormous crop losses. (In 1967 Kenya lost some $10 million worth of coffee from CBD.) (Shell photograph)

Plate 12.3. A young crop of sugar cane in Mauritius. A variety of insect pests attack sugar cane, and the potential annual loss, and gain from their control, is between one million and 16 million tonnes. (Shell photograph)

# 12 Some major tropical cash crops

The essential components of human diet are by no means the only annual crops that are important. People need more than food. For many countries, even those that can produce all the basic foods they need, the problem is to produce something that the rest of the world wants to buy. When they can do that, they can in turn buy, in the world's markets, goods that they cannot themselves produce.

Considering only the tropical developing countries, they can export minerals if these are available, but they are generally a wasting asset; or they can export crops that they can grow more readily than other countries. Some of these are foods, such as cane sugar, vegetable oils from a variety of plants, bananas, and ground nuts. Another group consists of the so-called stimulants, cocoa and the caffeine-containing beverages, coffee and tea. And there are rubber and the fibre crops, cotton, jute and sisal.

Cash crops have become highly important as exports from the less developed tropical countries and most of them are liable to pests and diseases so serious as to be economically disastrous if uncontrolled.

Cotton is so widely important and has so many pest problems that it is treated in a separate chapter. Each of the others chosen as examples is dealt with by an expert with many years of experience of the tropics, of the crop, and of the economic conditions and the people in some of the countries most concerned.

Some readers may be puzzled by the absence of precise figures of losses prevented by the use of pesticides. There are several reasons for this omission. First, it takes years of precise scientific work by specialists to obtain reasonably representative figures even with an annual crop grown on an English experimental station. Practical producers cannot afford such things. Account would also have to be taken of weather differences from year to year and from place to place, differences in soils, and differences of management that are often highly personal, in addition to differences arising from increasing age with the tree crops. In real-life agriculture, complexity is so great that precision could not be true and widely representative. Generally pesticides are used when their benefits are so obvious that no doubt remains.

Clearing land and planting it up for intensive production of crops is an adventure, formerly in little mapped territory, and the venturer used to meet many unforeseen hazards. Today, scientific knowledge provides help and warnings, warnings that go unheeded at the peril of the enterprise and the land. Even when there is no evident insuperable obstacle, there must be constant vigilance and ingenuity, characteristics without which man would have achieved little. And with established crops, vigilance and ingenuity are still required to see and solve new problems that arise.

# Rubber

## by C. C. Webster

Matching growth to consumer demand, world production of natural rubber from *Hevea brasiliensis* has increased steadily since 1945 to reach 3.5 million tonnes in 1974. Rubber is grown only in tropical developing countries and contributes significantly to the economies of Malaysia, Indonesia, Thailand, and Sri Lanka, which together produce 85 per cent of the world supply. In Malaysia, 500 000 smallholders produce 55 per cent of the current annual output of 1.5 million tonnes and altogether 3.5 million people derive their livelihood wholly or largely from the rubber industry, which accounts for 30 per cent of export earnings. Crop protection chemicals nowadays play an important part in achieving good yields and remunerative production.

Weed growth is limited by growing cover crops between the rows of young rubber and by the shade of the mature trees, but herbicides are extensively used, especially during the establishment of cover crops and to keep the tree rows free from weeds.

Because its latex provides a defence mechanism, rubber has no significant insect pests, but the hot, humid conditions and extensive plantations favour fungal diseases. When disease resistance is inadequate, fungicides are needed to control these diseases. The major root diseases, *Rigidosporus lignosus* and *Ganoderma pseudoferreum,* which can kill a third or more of the trees in a stand if unchecked, are mainly controlled by removing or isolating sources of infection, but fungicides are valuable as collar and root protective dressings on trees adjacent to diseased trees or to other sources of infection in the soil.

Fungicides are essential for the control of pink disease *(Corticum salmonicolor),* which can cause extensive loss of branches and delay maturity; and for the control of mouldy rot *(Ceratocystis fimbriata)* and black stripe *(Phytophthora palmivora* and *P.botryosa),* both of which damage the tapping panel and may prevent further tapping. Unless controlled early, all these diseases spread rapidly and are difficult to eradicate.

South American leaf blight *(Microcylus ulei),* which, in the absence of any known practicable treatment, has prevented plantation rubber production in the American tropics, does not occur elsewhere, but other foliage diseases often reduce growth and yield. Bird's eye spot *(Helminthosporium heveae)* is readily controlled by spraying, especially as it occurs mainly in nurseries. Mature leaf fall, caused by the two *Phytophthora* species mentioned above, is most important in India, where it necessitates annual spraying with copper fungicides to sustain good yields. The most widespread foliage malady is the repeated loss of the new leaves after the annual wintering, owing to *Oidium heveae* or to *Colletotrichum gloeosporoides. Oidium* can be controlled by suphur dusting during the period of refoliation, but spraying against *Colletotrichum* with currently available fungicides is less effective and of doubtful economic value.

Little quantitative information is available on the benefits of using fungicides. Their contribution to the control of root disease has not been distinguished from those of the other measures which must be simultaneously used. Fungicidal control of stem and panel diseases is essential, but the benefits have not been measured in terms of yield. The great variation in the severity of leaf diseases with local climate, clonal susceptibility and season makes it impossible to provide an overall estimate of the return from their treatment. Published experimental results record yield increases of up to 125 per cent from control of *Phytophthora,* but figures ranging from 15 to 50 per cent are more usual. Similarly, most reports indicate yield responses of 6 to 30 per cent from sulphur dusting against *Oidium,* but as much as 75 per cent has been obtained.

## Cocoa

### by E.E. Cheesman

The annual world production of cocoa (consumed mainly in the form of chocolate) is nearly 1.5 million tonnes, of which about two-thirds is grown in developing African countries. Ghana produces more than a quarter of the total and depends most heavily on cocoa for export earnings. It is also an important export from Nigeria, Ivory Coast, and Cameroon. Outside Africa the biggest producer is Brazil, and smaller quantities are grown by some half-dozen countries in the New World. In Asia and Oceania production is small, but increasing in New Guinea and Malaysia, where cocoa helps to provide a useful diversification of export crops.

The leading producers 'are all vitally concerned to preserve their markets and if possible to expand them; but, as chocolate is a luxury, demand is very sensitive to price. To maintain demand there must be a steady supply of cocoa at prices profitable to the growers but not high enough to discourage consumption. That ideal can only be attained by raising productivity.

Most of the crops of cocoa *(Theobroma cacao)* are grown by small farmers, and in general standards of cultivation, and therefore yields, are low. Research shows that yields can be increased spectacularly, sometimes by as much as fivefold or even more. Even a small fraction of such potential increase would enable world demand to be met more profitably from a smaller acreage, and so release a great deal of land for other crops including foodstuffs for the local populations.

Much of the higher productivity that must be achieved will accrue from the selection of higher-yielding trees and improved cultural methods; but a very substantial contribution can be expected from better crop protection. Cocoa has an impressive list of insect pests and fungal and virus diseases, favoured by the high rainfall and forest-like conditions under which it is grown. Because they vary in incidence and severity from country to country and from district to district, no precise estimate can be made of the total loss they cause; but it is certainly heavy and often obvious. One fungal disease alone (a pod rot caused by *Phytophthora palmivora)* is thought to cause a loss of crop amounting overall to some 10 per cent, rising in some localities to 80 per cent. Among the insect pests, capsid bugs, especially *Distantiella theobroma* and *Sahlbergella singularis,* have in the past caused even greater losses in West Africa (of the order of 50 per cent overall) and if uncontrolled could seriously damage the economies of several countries.

Under present conditions of low yield, protective measures known to be effective are often not economic. The *Phytophthora* pod rot has been the subject of spraying experiments in many countries over many years, yet spraying is not widely adopted in practice because too often the value of the crop that can be saved is less than the cost of spraying. Increased yields, however, will not only pay for better protective measures but will make them essential. More crop on the tree means more crop at risk before harvest, and more intensive cultivation will be useless without adequate protection.

As small farmers generally lack both capital and scientific knowledge, improvement in their techniques depends on assistance by their governments and especially the provision of efficient advisory services backed by research organizations. The governments concerned recognize these facts and have records of considerable successes in fighting some major pests and diseases on a national scale.

An outstanding example is the campaign directed against capsid bugs in Ghana, Nigeria, and French-speaking countries of West Africa since 1956, at which time crop loss from these pests in Ghana alone was estimated at 60 000 to 80 000 tonnes of dry cocoa annually. Following highly successful spraying experiments with gamma BHC, the government of Ghana organized a scheme under which more than 400 000 hectares of cocoa were sprayed to the end of 1958. Subsequently, mist blowers and insecticide were sold to farmers at subsidized prices. Between 1958 and 1961 Ghana's cocoa production doubled (from 220 000 to 450 000 tonnes) and it is generally agreed that capsid control measures were the main factor in this huge increase. Similar results were obtained in the other West African countries. The campaign continues to hold in check a most dangerous group of pests, although the situation has been complicated by the development of resistance to BHC by some capsids (see Chapter 16).

Control has thus been shown to be both possible and profitable even over the large areas of low-yielding cocoa that must be protected while yields are raised by other measures. Research has so far suggested no means of protection without the use of pesticides, which are likely to become increasingly necessary as heavier crops are concentrated in smaller areas.

**Further reading**

1  COLLINGWOOD, C.A. *Cocoa capsids in West Africa. Report of the International Research Team, 1965-71.* Cocoa, Chocolate and Confectionery Alliance, London (1971).
2  GREGORY, P.H. (ed). *Phytophthora disease of cocoa.* Longmans, London (1974).
3  THOROLD, C.A. *Diseases of Cocoa.* Clarendon Press, Oxford (1975).
4  WOOD, G.A.R. *Cocoa,* 3rd edn Longmans, London (1975).

# Tea

## by W. Wilson Mayne

Although tea is probably the most widely consumed non-alcoholic beverage in the world, its place in international trade is modest. Virtually all the tea produced in China, Japan, and the USSR and half that in India is consumed locally. Of the 688 000 tonnes of tea exported in 1972, India and Sri Lanka (formerly Ceylon) accounted for 400 000 tonnes. Tea dominates the export trade of Sri Lanka, earning nearly two-thirds of its foreign exchange, and is important to India in providing about one-sixth of its exports. Smaller amounts are exported from some 20 other countries, of which Indonesia and Kenya are the most important.

Commercial tea consists of young, immature shoots of *Camellia sinensis* treated too mildly to remove traces of protective chemicals. Consequently, chemical applications require exceptional discrimination to avoid toxicity and taint. The systematic use of pesticides in tea began only after 1946 when a fungal disease, blister blight, caused by *Exobasidium vexans,* first appeared in southern India and Sri Lanka and when DDT was first tested.

Mosquito blight, caused by the sucking of a capsid bug *(Helopeltis* spp. especially *H.theivora),* had earlier been investigated in India and DDT was immediately recognized as a possible control material; but the fungus of blister blight was new to tropical tea areas, though it had given trouble in north-east India. It spread rapidly throughout the 300 000 hectares of tea in southern India and Sri Lanka and appeared in Indonesia in 1949. The resemblance to the spread of coffee leaf rust *(Hemileia vastatrix)* in the 1870s, which eliminated commercial coffee production from Sri Lanka, was close enough to cause considerable alarm.

Copper fungicides were known to protect tea against blister blight and the problems of large-scale application were rapidly solved by the combined efforts of tea research institutes, agrochemical manufacturers, and planters. The innovation was very-low-volume spraying with fine suspension of the copper compounds. Protection required only 200-300 grams of copper per hectare at each of 15-25 applications per annum (i.e. annual amounts of only about 5 kg/ha).

These diseases were economically controlled with resultant crop increases of about 25-35 per cent. An uncontrolled attack of blister blight could wipe out the total crop for several months. Without these measures, tea in tropical Asia might have followed the pattern of coffee in Sri Lanka in the 1870s. The gains were not without complications, however, for the insecticide DDT and copper fungicides aggravated attacks of mites, and these now cause more concern than blister blight. The experience gained has been applied with appropriate pesticides to new problems and suitable acaricides are now used when necessary.

With shot-hole borer *(Xyleborus fornicatus),* a serious but localized invader of tea branches, there was transient success. Suitably timed applications of dieldrin were found in Sri Lanka in 1955 to give considerable control. Yield increases of 15-25 per cent were found on estates. When extended over larger continuous areas, however, the numbers of beneficial insects parasitizing other pests were reduced. Increasingly serious outbreaks of these pests led to the withdrawal of the recommendation of dieldrin for the control of shot-hole borer in 1966.

Such experiences are not unknown in pest control. Every plant provides food and shelter for a variety of organisms which normally live together in mutual toleration (see Chapter 20). Monoculture for mass production can upset this and some species become pests. Any control measure against a pest can alter the situation for other species and perhaps create new pests.

Of course, producers do not give up in the face of such complications. They search continually for methods that will favour the crop at the expense of pests and perhaps of neutral species. The applied scientists' role is to help in this search. Over the past century, they have extended the materials and techniques available so greatly that a catastrophe like the destruction of the Sri Lanka coffee industry is now unlikely.

## Further reading

1  CRANHAM, J.E. *Insect and mite pests of tea in Ceylon and their control. Monographs on Tea Production in Ceylon, no. 6.* Tea Research Institute, Talawakele (1966).
2  EDEN, T. *Tea* 2nd edn, Longmans Green, London (1965).
3  HAINSWORTH, E. *Tea pests and diseases.* Heffer, Cambridge (1952).
4  VENKATA RAM, C.S. *Blister blight of tea. Advances in Mycology and Plant Pathology.* India Phytopathological Society, New Delhi (1974).

# Coffee

## by J.M. Waller

Coffee is one of the world's most valuable agricultural crop commodities. World trade in it in 1973-4 was worth nearly $4000 million and was surpassed only by wheat, sugar, and oil seeds. Most of the trade deals with arabica coffee, from *Coffea arabica;* but robusta coffee, from *Coffea canephora,* is an important crop in some African countries. Coffee is a major export commodity and earner of foreign exchange for many tropical countries such as Brazil, Colombia, El Salvador, Costa Rica, Ivory Coast, Uganda, and Kenya.

As with many other crops, pests and diseases of coffee are most numerous within its centre of diversity, which is East and Central Africa; here, plant protection has always been an important aspect of coffee cultivation. In Asia and Latin America, however, in the absence of these constraints, coffee grew particularly well when it was first introduced; Latin America currently produces over 60 per cent of the world's coffee. The greater production away from their centres of diversity is a measure of the extent to which pests and diseases that evolved with these crop plants have restricted production in their original habitats.

Diseases and insect pests eventually spread to these new areas, and coffee rust *(Hemileia vastatrix)* was largely responsible for the destruction of the Sri Lanka coffee industry between 1870 and 1880. This is a classic example of the devastation of an unprotected crop by disease. Subsequently, rust severely restricted coffee production in Asia and Africa until chemical control, using copper fungicides, was established in the 1930s. More recently, the disease appeared in Brazil, thus threatening the whole of the Latin American coffee crop,[4] but fungicidal sprays are now achieving successful control there. Few of the many other leaf diseases of coffee warrant the expense of regular chemical control.

Probably the most directly destructive disease of coffee is coffee berry disease (CBD) caused by *Colletotrichum coffeanum.* This disease, which can cause virtually complete loss of crop on very susceptible trees, is at present confined to East Africa. Much effort was put into controlling CBD between 1955 and 1970, when it was causing financial distress to many farmers and threatened to ruin the best coffee areas in Kenya.

In 1967, a particularly bad year for CBD, at least one-quarter of Kenya's coffee crop, amounting to some 12 000 tonnes worth $10 million, was lost to the disease. Successful control measures were subsequently shown to depend upon critical placement and timing of fungicidal sprays.[2] These could treble the yields of coffee on well-managed estates for a cost-benefit ratio of at least 1:5. Chemical control of CBD and rust is now practised by most growers of arabica coffee in East Africa.

Varieties resistant to disease are little used commercially for several reasons. These include: lack of stability of resistance, a major problem preventing wider use of genetic resistance in many plants; the high cost of replanting perennial crops, including the loss of income before the new plants begin to yield·fully; and the need to maintain traditional quality which new varieties are unable to match. Nevertheless, work continues on producing stable resistance to rust in Brazil and varieties resistant to CBD are used in Ethiopia.

One of the most troublesome and widespread pests of coffee is leaf miner *(Leucoptera* spp*)*. Populations of these insects can quickly build up to devastating numbers causing defoliation and subsequent reduction of the crop. Chemical control of this pest with insecticides can double yields and is necessary in many countries for economic coffee production.

The berry borer, *Stephanodores (=Hypothenemus) hampei,* caused much damage when first introduced into Brazil in the 1930s. Chemical control is required both to prevent damage and to contain its spread within Latin America.

Of the many other pests that attack coffee in East Africa, most are adequately controlled by natural predators and by their own pests or by cultural measures. Nevertheless, the use of insecticides is often necessary to supplement these controls, for dealing with upsurges in pest populations, such as those of the leaf miner and of *Antestia,* which can also severely damage unprotected crops.

From these examples it can be seen that protection of the coffee crop, often requiring chemical means, has been and still is essential for the economic wellbeing of many tropical countries. However, combined methods of control have not been ignored and are continuously being developed and used (see Chapter 20).

**References and further reading**

1  FIRMAN, I.D. and WALLER, J.M. Coffee berry disease and other *Colletotrichum* diseases of coffee. *Phytopath. Pap.* **20** (in the press).
2  GRIFFITHS, E., GIBBS, J.N. and WALLER, J.M. Control of coffee berry disease. *Ann. appl. Biol.* **67**, 45–74 (1971).
3  LE PELLEY, R.H. *Pests of coffee.* Longmans, London (1968).
4  WALLER, J.M. Coffee rust in Latin America. PANS, **18**, 402-8 (1972).

# Cane Sugar

by F.K.E. Imrie

The world production of sugar in 1974 was about 81 million tonnes; of this, about 58 per cent was derived from sugar cane and the rest from sugar beet. Beet sugar is mostly consumed in or near the country of origin, but cane sugar is exported all over the world.

Sugar cane is a tropical and subtropical crop grown between 37°N and 35°S latitudes. In Mauritius, Barbados, and Cuba, sugar is a main source of foreign exchange; in other countries — Australia, the U.S.A., and India — much sugar is produced but its economic importance is less.

Many different insect species feed on sugar cane; they are mostly local insects that have adopted sugar as a host because it is there rather than because of any specific pest-host relationship. Thus, locusts, termites, and root-eating grubs damage sugar cane but are equally damaging to other plants. The most important pests are froghoppers and moth borers.

Froghoppers are common pests of the cane only in the New World, where 66 species are known. The nymphs may feed on either the roots or the stem; the adults feed on the leaves. Root feeding causes yellowing of the leaves and the entire growth may be stunted. Leaf feeding causes necrosis at each feeding puncture, which may ultimately destroy the entire leaf and slow down growth.

Losses caused by froghoppers may be large. Experiments in Trinidad showed that moderate damage reduced sugar yields by as much as 40 per cent and that losses exceeding 2.5 t/ha were common.

Froghoppers are major pests in Trinidad, Mexico, Venezuela, Belize, Brazil, and Guyana. The nymphs are controlled by dusting the bases of the cane plants with insecticides having long residual activity, such as gamma BHC, dieldrin, phorate, or toxaphene. Adults are controlled by aerial spraying of gamma BHC or DDT. Recently, foliar sprays of carbaryl or imidan have proved more effective. In Trinidad, where nymphs developed resistance to organochlorine compounds, organophosphate and carbamate insecticides were successfully substituted.

Moth borers have larvae that damage sugar cane. They tunnel into the stalk, causing extensive direct damage and also allowing secondary attack by fungi and bacteria to occur. Sugar cane of any age may be attacked and some cane may be killed, reducing the yield of cane. The cane may also be weakened and infected by micro-organisms such as gum-forming bacteria, so that the cane yields less sugar per tonne.

Early observations in Louisiana, when the borer moth *Diatraea saccharalis* was spreading through southern USA, showed that an overall loss of 1.33 t/ha of sugar occurred. In some countries, such as South Africa and Mauritius, natural enemies exert effective control over infestations. In Cuba, the tachinid predator *Lixophaga diatraeae* is artificially reared and liberated in cane fields as a control measure. Only in Louisiana are insecticides (endrin, guthion, carbaryl, and endosulfan) used to any great extent. More recently the systemic insecticide Azodrin has been used against endrin-resistant strains.

White grubs are the larvae of several families of beetles belonging to the superfamily Scarabaeoidea. Of 200 species associated with sugar cane, only a few are important pests. These can cause serious losses in Queensland, Puerto Rico, the Philippines, Mauritius, India, Hawaii, Taiwan, and East Africa.

The grubs feed on the roots, in severe attacks destroying them all, leaving the plant to die. Even a minor infestation will reduce sugar yields. In Queensland, losses over the period 1946-55 averaged over 13 t/ha of cane. In Puerto Rico reductions of 35-50 t/ha were being recorded as recently as 1967. Some countries, notably Mauritius and Hawaii, have been very successful with biological methods of controlling the grubs. In other areas, persistent soil insecticides such as BHC have been effective.

Termites of some 53 species have been recorded as feeding on sugar cane. Crop losses are usually minor and merely a nuisance, but in Bihar,

India, termites are a major pest. Up to 60 per cent of buds on newly planted setts may be destroyed, causing a loss of up to 5 per cent of sugar yield. In Taiwan, a species of *Coptotermes* is a major pest. Persistent organochlorine insecticides — aldrin, dieldrin, and chlordane — are effective.

### Further reading

WILLIAMS, J.R., METCALFE, J.R., MUNGOMERY, R.W., and MATHES, R. (eds). *Pests of sugar cane.* Elsevier Publishing Co., Amsterdam. (1969).

*                    *                    *

The accounts given above show clearly how impossible it is to estimate accurately potential losses with no control, or the value of applying control to pests. All concerned are convinced of the need for rational measures of control. With cane sugar the potential loss and the gain from its prevention lies somewhere between one and 16 million tonnes of sugar annually.

# 13 Cotton

by G.A. Matthews

Most people think of cotton as a textile fibre, as indeed it is, but about two-thirds by weight of the seed cotton harvested is seed. When the seed is crushed a valuable vegetable oil is extracted and the remaining cake is rich in protein (Fig. 13.1).

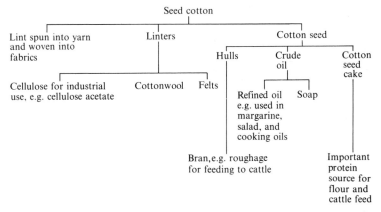

Fig. 13.1. Some uses of cotton products. Based on an abbreviated version of the table of cotton seed products in ref. (1), p.42.

Despite strong competition from synthetic fibres, the world-wide demand for cotton is increasing. To help cotton compete successfully various agricultural inputs, including pesticides, are needed to optimize production. Cotton is grown in over 60 countries on a total of about 32 000 000 hectares. This represents about 2.5 per cent of all cultivated land. Total production was estimated at 13 612 000 metric tonnes of lint (fibre) in 1974-5, which, at approximately 50p/kg,[1] has a total value of £6 806 000 000.

The economic importance of cotton in the agriculture of many countries is shown in Table 13.1. The crop is particularly important in the U.S.A., China, and the U.S.S.R., but in proportion to total agricultural

---

[1] The lint price of 50p/kg is based on the Liverpool Spot price; as this fluctuates it is not possible to discuss prices in any detail.

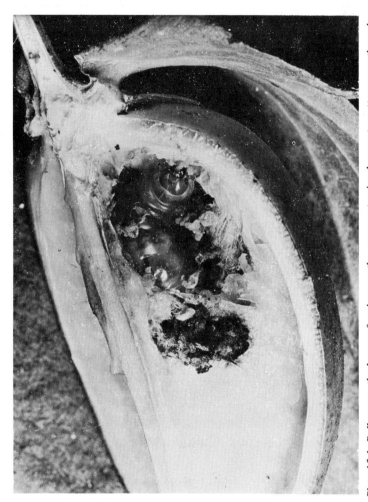

Plate 13.1. Bollworms, the larvae of various moths, cause extensive damage to cotton crops throughout the world. This photograph, from Venezuela, shows the bollworm *Sacadodes pyralis* Dyar, feeding on a cotton boll. (Shell photograph)

Table 13.1. *Estimated lint production and yields 1974-5 with exports and imports 1973-4*

|  | Prod. 1974-5 (000 tonnes) | Yield 1974-5 (kg/ha) | Exports 1973-4 (000 tonnes) | Imports 1973-4 (000 tonnes) |
|---|---|---|---|---|
| Central African Republic | 17 | 128 | 15.2 | — |
| Chad | 50 | 184 | 39 | — |
| Egypt | 434 | 712 | 261.1 | — |
| Nigeria | 65 | 154 | 1.7 | 3.7 |
| Sudan | 206 | 424 | 170 | — |
| Tanzania | 76 | 268 | 65 | — |
| AFRICA: total/mean | 1207 | 294 | 824.1 | 86.1 |
| Mexico | 477 | 841 | 159.8 | — |
| U.S.A. | 2537 | 497 | 1333.2 | 10.4 |
| NORTH AMERICA: total/mean | 3285 | 542 | 1723.4 | 107.5 |
| Brazil | 499 | 224 | 143.8 | — |
| Columbia | 150 | 521 | 43.4 | 6.5 |
| SOUTH AMERICA: total/mean | 958 | 278 | 280.7 | 58.7 |
| Australia | 39 | 1134 | 1.3 | 4.6 |
| China | 2147 | 446 | — | 412.0 |
| India | 1193 | 155 | 39 | 27.1 |
| Pakistan | 640 | 351 | 47.5 | 0.4 |
| Turkey | 585 | 804 | 217.7 | — |
| ASIA/OCEANIA: total/mean | 5122 | 317 | 566.6 | 2116.5 |
| U.S.S.R. | 2840 | 1003 | 758.9 | 130.1 |
| EUROPE WITH U.S.S.R.: total/mean | 3040 |  | 1088.6 | 1962.6 |
| WORLD | 13616 | 414 | 4202.7 | 4331.4 |

Source: *Quarterly Bulletin of the International Cotton Advisory Committee*, April 1975.

*Note;* Certain countries such as Egypt and Sudan produce a long-staple cotton, *Gossypium barbadense,* whereas the principal cotton grown elsewhere is *Gossypium hirsutum.*

production it is probably more important in many of the developing countries of Africa, Asia, and Latin America. Many of the developing countries export most of their cotton to earn foreign exchange; for example, Tanzania, which exports over 80 per cent of its cotton, earns at present-day prices about £30 000 000. The Sudan is another example of a country which exports 80 per cent of its cotton. On average, two-thirds of the cotton grown in Africa is exported. However, some countries have established textile factories. Among these, Nigeria is currently importing cotton to meet local demand.

Under peasant farming conditions, increased production was often achieved by farmers using a larger area rather than raising the yield per hectare of cotton. Even with careful cultivation to control weeds, yields of unprotected crops are reduced by insects and diseases to under 500 kg/ha seed cotton (160-80 kg/ha lint). When a farmer is paid 12p/kg for his clean seed cotton (approximately equivalent to 36p/kg for lint on the farm), he would receive under £60/ha for his crop. As about 150 man-days are required to prepare one hectare of land, sow, weed, pick, grade, and market the crop, his income would be only about 40p per man-day.

Yields of rain-grown cotton can be doubled by applying pesticides. Even higher yields are possible on protected crops if the rains are favourable. After deducting the cost of spraying, which is estimated at up to £30/ha for 12 insecticide sprays, the farmer's income is increased, but he has to spend more time on spraying and picking a larger crop. In irrigated areas, protected crops have yielded in excess of 2500 kg/ha seed cotton over large areas.

Before the introduction of organic insecticides, cotton farmers had to rely on biological and cultural techniques to reduce damage by the many insect pests attacking the crop from sowing to harvesting. Control was attempted by manipulating sowing dates, releasing parasites, and using trap crops, but yields still remained low. To understand the importance of the various insect species we must consider the growth of cotton plants.

### The cotton plant

Conditions for cotton growing vary enormously. To mention some extremes, the crop is grown in areas of low rainfall under irrigation, as in the Sudan, in wet equatorial areas, as in parts of Uganda, and also in drier savannah areas. Except where irrigation permits some control of sowing date, the seed is usually sown at the start of the rains or when any risk of cold weather is over. The plant produces a main stem usually with one or more vegetative branches according to the space between plants. Successive fruiting branches develop along the main stem and the vegetative branches.

The first flower buds appear about six weeks after germination and break into flower for a day about two weeks later. Subsequently they form the first fruit or cotton boll. Bolls take 35-70 days to mature according to local conditions. An excess of flower buds is produced (sometimes over 100 per plant), so that even plants which yield well shed large numbers of buds or small bolls without insect or disease damage.

Water stress and falling temperatures are important towards the end of the season and cause considerable late shedding. The farmer needs to protect the earliest-formed buds artificially to allow them time to mature and produce and best-quality cotton. Correct timing of pest control is therefore of paramount importance; so ideally pest populations need to be monitored to determine economic thresholds, namely the maximum pest population that can be tolerated without resultant crop loss. But first the farmer must provide conditions for optimum plant growth so that bud production is not delayed. This involves timely sowing with a disease-resistant variety where possible. High-quality seed should be used at a suitable plant density and spacing.

If fertilizer is needed, great care must be taken to ensure a proper balance of nutrients, including certain trace elements such as boron. Excessive nitrogen promotes lush growth attractive to insect pests and it makes chemical control difficult. Cotton is a deep-rooted crop and it can therefore survive quite prolonged drought periods, remaining green and attractive to insect pests over a very long period, in contrast to crops such as maize. In the drier parts of the tropics, survival of a cash crop like cotton enables the farmer to purchase food if his maize crop fails.

## Plant establishment

The ability of cotton to compensate for losses of seedlings has been recognized for a long time, for similar yields are obtained over a wide range of plant densities. There is, however, an optimum at which bud formation and boll ripening proceed fastest, thus reducing the period over which crop protection is required. High yields from widely spaced plants can be achieved only by waiting for the later bolls to open at the ends of branches. The onset of cooler weather at the end of the season can delay ripening of these late bolls considerably and diminish the quality of the lint, apart from exposing them to insect attack for longer. Conversely, too dense a plant stand has disadvantages similar to an excess of fertilizer, and may restrict penetration of light and air, thus reducing photosynthesis and increasing natural shedding of buds or rotting of bolls.

In many African countries, to obtain a good stand where early season losses were caused by termites and various grasshoppers, farmers frequently sowed surplus seed and thinned seedlings about three weeks later, although both seed and labour were wasted. The trend is to use less seed but of a higher quality, delinted and treated with pesticide to protect young seedlings from pathogens and insects.

### Some of the major insect pests

An exceptionally wide range of pests attack cotton plants at all stages of growth. Early in the season the sucking pests, jassids *(Empoascu* spp*)* cause leaves to curl, redden and eventually shed. Their damage was so serious in parts of Africa that the cotton industry was saved in the 1920s only by the introduction of a resistant variety with hairy leaves. Where cotton is harvested by machine, as in the U.S.A., smooth-leaved, high-yielding varieties are preferred to reduce the amount of leaf trash clinging to the lint. These high-yielding varieties have been sown in other parts of the world, but with disastrous results where jassids occur, except where adequate plant protection measures were taken.

Jassid infestation in some seasons can develop so rapidly, particularly in hot, dry areas, that plants even of resistant varieties can be stunted. In this case the use of an insecticide is justified. The choice of chemical is important: thus, in the Sudan, the use of DDT sprays resulted in an increase in population of another leaf-sucking insect, whitefly *(Bemisia tabasci)*, and other insecticides had to be used.

Natural enemies of pests tend to succumb to insecticides but the whitefly nymphs are well protected on the undersides of leaves, especially on the lower parts of plants. Even systemic insecticides are not readily translocated to these leaves. Control of nymphs as well as adults is essential over an extensive area, otherwise the pest population will rapidly re-establish from freshly emerging flies, and also by immigration, for they are carried long distances by the wind. Infestations of whitefly are particularly important on long staple cotton varieties as a vector of the virus disease, leaf curl. On all cotton varieties, the spinning quality of the lint is affected by stickiness caused by secretions of honey-dew on which sooty moulds may develop. Control of whitefly late in the season is therefore essential to avoid downgrading of lint quality.

World-wide, the most extensive damage to buds, flowers, and bolls is caused by the feeding of the larval stage of various moths. These larvae are called bollworms. The main genus throughout the world is *Heliothis*, which feeds on a wide range of plants during flowering. Control of pests on cotton early in the season permits more flowers to open and thus attracts heavier infestations of *Heliothis*. Damage is also more extensive when alternative hosts are grown in sequence throughout the year with irrigation. *Heliothis* was not a serious pest in the Sudan Gezira until fallows were replaced by groundnuts and sorghum, and the time of sowing cotton was changed. Total loss of crop can occur if *Heliothis* is not controlled, for each larva can damage up to 15 buds or bolls.

When moths migrate into cotton fields, populations can increase so rapidly over large areas that natural parasite and predator populations are unable to control infestations. The main problem, therefore, is to detect changes in numbers of eggs so that as soon as the economic threshold is reached, insecticides can be applied quickly to reduce the number of first-stage larvae entering a bud or boll. Bracts protect larvae while feeding, although plant breeders have tried to introduce narrow, twisted ('frego') bracts and 'okra' leaves on commercial varieties to expose larvae to insecticide sprays. As eggs hatch in two to three days, control measures have to be put into effect quickly. Populations must therefore, be monitored by routine inspection (scouting) for eggs, which are mostly laid in the upper half of the plant.

In some areas, spray applications are timed according to counts of larvae or damaged buds, but unless threshold values are set low, extensive damage may occur before sprays are applied; and larger larvae are far more difficult to kill. Unfortunately, oviposition continues over a period, and so several sprays are necessary to protect new buds. A lethal deposit on surfaces of the plant where moths and larvae walk is required for efficient insecticide application. So often sprays are directed down over the crop and are filtered out, mostly on the upper leaves. Local over-dosing and 'run-off' of spray occurs, while larvae are protected by the foliage and continue to feed. Often a farmer applies pesticides as quickly as possible without considering how much chemical reaches the correct target. When control breaks down, he may apply another, perhaps larger dose generally to the same parts of the plants, without seeking improved distribution of the chemical.

Improved control will depend on selection of the most suitable droplet size to obtain maximum penetration and collection of droplets within the plant framework. Control of *Heliothis* has been achieved with the application of 2-3 l/ha with very small droplets (70-120 $\mu$m diameter) of a relatively non-volatile formulation of insecticide. Droplets were produced with a spinning disc nozzle held above the plants and the droplets were deposited on the plants as a result of gravity and natural air movement. Further work on this type of approach to insecticide application is needed.

Plant breeders are seeking to improve plant resistance to bollworms by selecting varieties with such factors as higher gossypol content. There are tremendous difficulties, however, for plant breeders who must compromise between the need to achieve resistance and yet maintain the quality and yields required by the textile industry.

In the U.S.A., Mexico, and Central America, in addition to bollworm damage, boll-weevils cause heavy losses (estimated at over $ 200 million each year in the U.S.A.). Unfortunately, use of certain broad-spectrum insecticides can result in an increase in red spider mite infestations. Mites cause premature defoliation of the crop with subsequent reductions in the yield and quality of seed cotton. Mite resistance to a wide range of acaricides has developed very rapidly. To delay resistance, systems of rotating the use of different chemical groups are used in some countries to prevent simultaneous resistance to several groups occurring on neighbouring farms.

Late in the season a crop may be 'red with stainers'. When the bolls are splitting open, cotton stainer damage to lint is minimal, but their feeding also damages the seed by lowering viability and reducing the oil content. Earlier, very few cotton stainer bugs, migrating from alternative hosts, can cause considerable damage if a fungus which stains and weakens the lint is transmitted at each feeding puncture. Stainers are often controlled by sprays applied for bollworm control. Without stainer control, up to 30 per cent of the lint is stained and the farmer has to accept a lower price, or spend hours sorting out the clean from the damaged lint. Also, excessively large stocks of seed have to be retained for sowing the next crop instead of sending it to the oil extractors.

Pink bollworm has spread to almost every cotton-growing area. Control with insecticides has not always been very successful, and so many countries rely on a closed season of at least three months to reduce the survival of larvae from one season to the next. As end-of-season cotton has a lower quality, it is better in the long term to sacrifice some of the crop to reduce the number of pests entering diapause. When farmers have failed to uproot their plants and burn the debris, or have deliberately pruned back the bushes to grow again in the next year and save sowing seed, pest and disease problems increase, and young plants are infested severely before they can become established.

### Diseases

Compared with insect pests, plant diseases are not a major problem in cotton; where they occur, the most effective control of such diseases as fusarium wilt has been achieved by sowing resistant varieties. Some diseases, including bacterial blight, are seed-borne; seed is therefore treated with fungicides or is acid-delinted before sowing.

### Weeds

During the first six weeks of growth, weed control is absolutely essential to ensure a potential yield which is worth protecting against insect pests. Delay in weeding can reduce yields by as much as 450 kg/ha seed cotton, the sale of which would more than cover costs of insect control. Unfortunately, many peasant cotton farmers in the tropics rely on hand hoeing for weeding and are unable to cope with the rapid weed growth at the start of the season, when cotton foliage is growing slowly and the inter-row space is not covered. At this time, priority has to be given to food crops and extra labour for weeding in cotton during the short but critical period is seldom available.

Farmers who spray insecticides are advised to sow cotton only over an area which they can weed adequately; but this advice is seldom accepted, with the result that the economic return on investment for insect control is reduced. Herbicides have not been used widely by these farmers, for they find it impossible to collect the recommended volume of water, particularly at the start of the rains. Such farmers may have the use of a borehole in a near-by village, but often they rely on seasonal streams for supplies of water for spraying. Even when water is available, its transport to the fields is very time-consuming. The application of herbicides in as little water as 10 l/ha is, however, now possible with the recent development of equipment with which droplet size can be controlled. Efficient distribution of suitable herbicides by controlled droplet application may revolutionize weed control, not only in cotton but many other crops; but further research is needed to develop the technique under a wide range of conditions.

*                *                *

Unfortunately, farmers who use pesticides frequently assume that if they use more chemicals yields will continue to increase; but the lessons from disasters in cotton production in Peru and several Central American countries must be learnt before it is too late. For a variety of reasons, in the 1960s the number of spray applications increased to about 30 per season, but subsequent government action had reduced the number to around 20 by the early 1970s. Government action has also resulted in a marked reduction in the large number of cases of poisoning and death which had been occurring in Nicaragua, where highly toxic insecticides had been in use.

Apart from the unacceptable health hazard, the use of insecticides indiscriminately and on a large scale can result in unwanted effects such as contamination and pesticide resistance in other insects (mosquitoes, for example) together with lower yields of cotton. Cotton production then becomes uneconomic — as it has been in part of Mexico — and if farmers cease to grow the crop, the whole cotton marketing and manufacturing industry of the area is seriously affected, causing unemployment in that industry.

Gross over-use of chemicals is totally unnecessary if the farmer makes full use of resistant varieties and recommended cultural practices and then restricts the use of insecticides to what is strictly necessary. One of the main difficulties will be in educating farmers to consider the long-term economics of cotton production rather than to seek immediate maximum yields.

The successful use of pesticides in a pest-management programme requires careful selection of the most appropriate chemical for a given pest, monitoring of the pest population regularly to ensure correct timing of application in relation to economic thresholds, and improved application techniques. Further research is needed on the usefulness of insect sex attractants (pheromones) in monitoring insect populations to indicate the correct timing of insecticide applications, and also in controlling insects by disrupting their behaviour (see Chapter 20). Although the use of non-pesticide methods may increase, at present pesticides are needed. Solutions to the many complex pest problems of cotton growing will not be simple or cheap. With careful observation

and greater understanding of the problems, the intelligent use of
pesticides will allow the present level of production to be maintained
with far less manpower and resources.

### Further reading

1. ANON. *Cotton Technical Monograph 3,* Ciba-Geigy, Basle (1972).
2. FALCON, L.A. and SMITH, R.F. *Guidelines for integrated control of cotton pests.* FAO, Rome (1973).
3. MATTHEWS, F.A. Ultra-low volume spray application on cotton in Malawi. *PANS,* **19,** 48-53 (1973).
4. PEARSON, E.O. and MAXWELL DARLING, R.S. *The insect pests of cotton in tropical Africa.* Cotton Research Corporation and Commonwealth Institute of Entomology, London (1958).
5. PRENTICE, A.N. *Cotton.* Longmans, London (1972).
6. PROCTOR, J.H. A review of cotton entomology. *Outl. Agric.* **8,** 15-22, (1974).
7. RABB, R.L. and GUTHRIE, F.E. (Ed.) *Concepts in pest management.* North Carolina State University Press (1970).

Plate 14.1. A striking example of damage caused by the Khapra beetle (*Trogoderma granarium*) to shelled groundnuts in store in Senegal. The bags had burst as a result of the insect's activity. Attack by this beetle can cause an increase in free fatty acid content of the groundnuts, and this reduces the quantity and quality of oil. (FAO photograph by I. Pattinson)

# 14 Food in store

by J. A. Freeman

Settled agriculture has always required storage of crops and seed from one season to the next. Drying of crops and livestock products prevents rapid deterioration from attack by bacteria; dry products are, however, still subject to attack by storage insects and mites, by some fungi, and by rats, mice, and birds. These pests destroy millions of tons of grain and other agricultural products annually. The greatest direct losses in weight and quality occur in the damp tropics and the least in developed countries with temperate climates. But in the latter there may be heavy financial losses where the strict standards of hygiene imposed by public authorities in the manufacture and distribution of food are disregarded. Most of these losses could be prevented by the application of existing knowledge of hygienic storage and processing and the use of pesticides and other control methods.

## Insect and mite pests

The main insect pests are small beetles and moths, whose life-cycles are similar. The fertilized female lays an egg which develops into a larva; this moults several times, increasing in size at each moult; eventually the pupal stage is reached and change to the adult form takes place. The larva may be protected inside the endosperm of cereal grains (e.g., maize weevil, *Sitophilus zeamais*); in the germ (e.g., saw-toothed grain beetle, *Oryzaephilus surinamensis*); or in the cotyledon of the cocoa bean (e.g., the cigarette beetle, *Lasioderma serricorne*). It may feed on the surface of the food (e.g., the Indian meal moth, *Plodia interpunctella* on dried fruits and nuts), or between the particles (e.g., the rust-red flour beetle, *Tribolium castaneum,* or the Mediterranean flour moth, *Ephestia kuehniella,* in flour). Most beetles damage foodstuffs both as larvae and adults; but only the larvae of moths are destructive, for the adults feed on liquids.

Mites have a similar life-cycle but without a pupal stage; they also differ in being wingless in the adult stage. They live in particulate materials like flour, inside dried fruits, or in the germ of grain.

The nature and scale of losses in the tropics are striking. For example, parboiled rice stored for 12 months in Sierra Leone lost 41

169

per cent of its weight (after screening) from attack by the lesser grain borer *(Rhizopertha dominica);* natural paddy (i.e., the grain in the husk as it comes from the plant), was virtually undamaged. Some 4 per cent of the sorghum crop, sufficient to feed a million people, is lost each year in northern Nigeria, mainly to the rice weevil *(Sitophilus oryzae)* and the Angoumois grain moth *(Sitotroga cerealella).* Again, in northern Nigeria, by the end of the season cowpeas on sale are 50 per cent holed and have lost 25 per cent of their weight to Bruchid beetles. Because the cowpeas are sold by volume, not weight, the cost of such infested produce is very much greater than for sound cowpeas. Another example: some 50 per cent of the protein of smoke-dried fish from Lake Chad is destroyed annually by the leather beetle *(Dermestes maculatus)* during the 6 months between drying and sale in the markets of Kano, Ibadan, and Lagos in Nigeria. And when groundnuts are stored, attack by the groundnut seed beetle *(Caryedon serratus),* the rust-red flour beetle, and the khapra beetle *(Trogoderma granarium)* can cause an increase in free fatty acid content which reduces the quality and quantity of oil.

In temperate countries some trouble occurs because of the continued development (especially in heated buildings) of insects imported in products from warmer countries. Thus, imported dried fruit and nuts, although not losing much weight, can be fouled by silk webbing and droppings of tropical warehouse moth caterpillars, and by beetles. As yet one more example, two cargoes of dates were held up in London in 1975 while port health inspectors supervised the sorting and cleaning of the cargoes. In the end a high proportion of the shipment had to go for animal feed.

There are many species of pest insects and mites that are adapted to temperate conditions and can flourish in unheated warehouses in the temperate zone. The rust-red grain beetle *(Cryptoleste ferrugineus)* is a serious pest of stored wheat in Canada up to the northern limit of wheat growing. Again, mites can cause trouble by eating the germ of grain (especially the flour mite, *Acarus siro)* and by attacking cheese and dried fruit (especially the dried fruit mite, *Carpoglyphus lactis)* and also by reducing the nutritive value of animal feed. They may also cause allergies in man. In addition to contaminating foods by their presence and by their droppings, insects cause particular damage to grain by tunnelling into or by eating from outside the germ (warehouse moth caterpillars, *Ephestia elutella*), thereby reducing food value. During the Second World War these larvae destroyed 1.5t of germ in 500t of wheat over a period of two years.

Apart from direct damage, insects can cause heating in grain and other commodities stored in bulk. Spontaneous heating caused by an increase of insect population within a bulk results in the insects eventually being driven to the periphery, because the temperature within rises to a lethal level – normally a maximum of 42 °C. Such 'dry grain' heating may result in the movement of moisture from the hot interior to the periphery to such an extent that fungi can attack the grain and sprouting can take place at the surface. Fungal heating may raise the temperature to 65 °C. In this way a small initial infestation can result in a useless, mouldy, and sprouted mass.

Damage of these kinds may result not only in less grain being available for the original intended use, but also in diversion to some other less profitable use. Wheat which is too heavily infested by insects or mites or contaminated by rodent pellets may produce a flour with too high a count of microscopic fragments to be acceptable in some countries. Similarly, de-germed grain cannot be used for seed or malting, nor can insect-chewed almonds be used to decorate cakes.

## International trade

The actual damage resulting from infestation, or the potential damage represented by the presence of insects or mites, is recognized in national and international trade by clauses in commercial contracts concerned with transactions in grain, pulses, etc., and in the various official grading regulations of exporting countries. These place limits on the amount of insect damage which will be accepted, and may or may not tolerate the presence of living insects or mites. Thus, the Canadian and Australian regulations accept no tolerance for living insects in wheat for export; while the U.S.A. allows one live weevil or a larger number of other insects per kilogram of wheat or maize. Similar official regulations apply to cocoa beans exported from Nigeria and Ghana and to groundnuts from South Africa and the U.S.A. Ships and dry cargo containers into which such commodities are loaded are also officially inspected, and if they are infested they must be treated before loading is permitted. In this way clean commodities are protected against infestation remaining from previous voyages, and from insects and mites from other commodities.

Even so, damage can occur during voyages from the development of insects that cause heating, which is followed by fungal attack. Similar damage may result in dry cargo containers from condensation of moisture. The mere presence of insects can result in severe financial loss if an importing country imposes strict phytosanitary or quality standards. This is so for China, where the discovery of one living insect in a cargo of wheat can result in the shipment being rejected or accepted only after expensive fumigation.

## Mycotoxins; rats, mice, and birds

The presence of mycotoxins resulting from the attack by certain species of fungi on a product is now recognized as a serious threat to human and animal health. Aflatoxin, for instance, may be produced in groundnuts and brazil nuts, as well as in grain, by *Aspergillus flavus*, normally at temperatures over 25 °C. Other mycotoxins have been found in grain stored under temperate conditions. Food and feeding-stuffs manufacturers and public health authorities test susceptible products before use to ensure that the finished products do not contain dangerous levels of these organisms.

It must also be noted that rats, mice, and birds can cause damage by eating and fouling commodities with droppings and urine, and by damaging sacks. And their nests and droppings can harbour storage insects and mites.

## How infestation starts and develops

In temperate countries, infestation of grain and other commodities by storage insects and mites starts once crops have been placed in store; in the tropics, infestation may occur in the ripening grain or pulse crop. This is especially true of attacks on maize by the maize weevil and the Angoumois grain moth, which fly into the fields from near-by grain stores. Hybrid maize with cobs longer than the sheath leaves is particularly liable to attack; so are cobs already damaged by field-crop pests. The closer the crop is to the store, the higher will be the incidence of infestation.

As storage insects and mites usually have plenty of food, their increase in numbers is normally determined by temperature and relative humidity. Most storage insects are active within the range 10 - 40 °C; mites have a lower threshold, increasing slowly in numbers down to 0 °C. Both prefer relative humidities higher than 60 per cent, corresponding, in dry wheat, to a moisture content of about 13.5 per cent. In general, different species of insects and mites are adapted to different ranges of temperature and relative humidity (or moisture

content), and their numbers increase under clearly defined limiting
and optimum conditions.

So far as grain is concerned, the khapra beetle is the principal pest
under very hot and dry conditions; as temperature falls and moisture
content increases this species is replaced by the lesser grain borer, then
by the rice weevil, then by the grain weevil *(Sitophilus granarius)* and
finally by the flour mite. For grain to be safe against attack by fungi
(mainly *Penicillium* spp and *Aspergillus* spp), commodity moisture
content needs to be in equilibrium with air of 70 per cent relative
humidity or less, although higher levels can be tolerated at low
temperatures. Thus, the safe maximum moisture content for storage of
wheat under temperate conditions (say 15 °C) without artificial

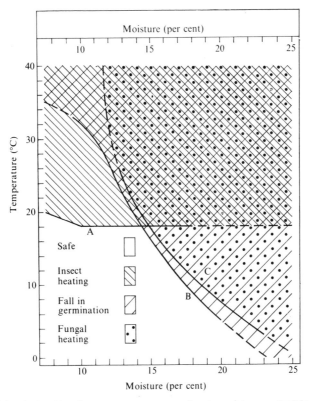

Fig. 14.1 Relationship of storage temperature and grain moisture content to insect
heating reduction in germination and damp grain heating. (Source:
ref. (1).) *Crown copyright.*

cooling, for periods of storage over 6 months, is 14 per cent, and correspondingly less in the tropics.

The limiting conditions for safe storage are indicated in Fig. 14.1. They can be applied not only to grain but to other commodities, taking into account the different moisture content - relative humidity equilibria.

## Prevention and control

Since the stored crop, unlike the growing plant, cannot compensate for pest attack, every gram lost to pests represents an irreplaceable loss of food and energy. Although control measures may arrest further damage, they cannot replace what the pest has consumed, so prevention of attack is all-important. The basis of prevention is good hygiene, which implies that all storage places and machinery used for storage, transport and processing of food and animal feed stuffs must be free from places where residues accumulate and must be easy to clean. They should also be proof against rats, mice, and birds.

Hygiene alone, however, is not sufficient. Chemicals, used as contact insecticides and acaricides and fumigants, are necessary for controlling residual infestations, for prophylaxis, and for dealing with outbreaks. They may also be used against rats, mice, and some birds. In most countries the use of pesticides on or near food is strictly regulated in the interests of public health and the protection of wildlife. In deciding on the ways in which pesticides may be used, national Governments take into account the recommendations of the WHO-FAO Codex Alimentarius Commission regarding maximum residue levels (see Chapter 19). In Great Britain and elsewhere the use of pesticides is regulated by agreement between Government and industry under the Pesticides Safety Precautions Scheme, which lays down the conditions under which pesticides may be safely used (see Table 14.1).

Residual contact insecticides and acaricides applied as dusts, oil emulsions, or water-dispersible sprays, as oil sprays or as aerosols (smokes) form persistent films of materials toxic to insects and mites which crawl over or alight on treated surfaces or commodities. The main chemicals used are malathion, malathion-lindane mixture, pirimiphos methyl, fenitrothion, and dichlorvos and natural and synthetic pyrethrins. Propionic acid is used to inhibit growth of bacteria and fungi on high-moisture grain intended for use as animal feed.

Fumigant gases act by penetrating into commodities and into insects and mites. There is no effect after the gas has dispersed. Commodities

Table 14.1 *Some contact insecticides, acaricides, and fumigants,*
*(updated from ref. (4))*

| Type and British Standard name | Codex tolerances (ppm) | Food storage insects | Food storage mites‡ |
|---|---|---|---|
| **CONTACT INSECTICIDES AND ACARICIDES** | | | |
| *Organochlorine compounds* | | | |
| Gamma-BHC (lindane) | 0.5 | XM (>2.5 ppm) | XM |
| DDT† | – | X | |
| Chlordane | – | X | |
| *Organophosphorus compounds* | | | |
| Bromophos | 0.2 | XM (>12 ppm) | |
| Dichlorvos | 2.0 | X | X |
| Fenitrothion | – | X | X |
| Iodofenphos | – | X | X |
| Malathion | 8.0 | XM (>10 ppm) | XM |
| Phoxim | – | X | XX |
| Pirimiphos methyl | – | XM (>4 ppm) | XXM |
| Malathion + lindane | – | XM (>10 + 2.5 ppm) | XXM |
| *Pyrethroids* | | | |
| Pyrethrins (or synthetic pyrethrins) | 3.0 | | |
| + piperonyl butoxide | 20.0 | XM | |
| **FUMIGANTS** | | | |
| Hydrogen phosphide | 0.1 | X | X |
| Methyl bromide | 50 (as inorganic bromide) | X | XX |

The contact insecticides, acaricides, and fumigants listed are used against food
storage pests in Great Britain and are cleared under the Pesticides Safety
Precautions Scheme,* together with tolerances for residues in raw cereals
recommended by the Codex Committee on Pesticide Residues in 1975.

* Limitations in use under the scheme are given on labels.
† Only when other materials are not suitable.
‡ Susceptibility varies from species to species; see ref. (8).
XX materials to which *Acarus siro* and *Tyrophagus putrescentiae* are both
susceptible.
X indicates general use of pesticide: only those marked M may be mixed
with grain (limit shown in ppm).

to be fumigated must either be enclosed in gas-tight silos or special chambers, or else covered with gas-proof sheets. Fumigants in common use include methyl bromide, hydrogen phosphide (applied as proprietary mixtures of aluminium phosphide and other chemicals which combine with air moisture to produce phosphine), and mixtures of carbon tetrachloride, ethylene dichloride, and ethylene dibromide. Dosages and periods of exposure of fumigants are based on experimental work in laboratory and field, designed to ensure that the concentration-time product is adequate to kill the insect or mite without leaving an excessive or harmful residue in the product treated.

### Control of rats, mice, and birds

Despite the most careful attention to proofing, rats and mice eventually get into storage buildings; they may be controlled most economically by direct baiting with anti-coagulant drugs, but the appearance of resistance to the most widely used material, Warfarin, has made it necessary to seek replacements. Of these, mixtures of warfarin and calciferol (vitamin $D_2$) appear to be the most successful. Warfarin-resistant mice may be killed by alpha-chloralose baits, which cause hypothermia.

In many countries there are severe legislative restrictions on the killing of birds, but one permitted technique for birds entering grain stores, bakeries, etc., is to use a narcotizing bait. Pest birds may then be picked up and killed, and protected birds allowed to recover and escape.

### Control of insects and mites by non-chemical means

As many storage insects increase in numbers only slowly below about $12 \, °C$, keeping commodities cool is a good method of control, and it does not entail the use of pesticides. Valuable commodities like tobacco, dried fruit, and nuts can be protected against insects by keeping them in commercial cool stores; and grain stored in temperate countries can be mechanically ventilated with unheated air, whenever temperatures and relative humidities are low enough. Refrigerated air can be used for grain of moisture content up to 20 per cent. Flour mills in cold countries can be disinfested by 'freezing-out' in mid-winter.

Another possibility is to irradiate with sterilizing dosages (16 000 rad) of gamma rays from a cobalt-60 source, or by electrons from an electron accelerator. This technique may be used without ill effect on grain, but at present it is uneconomic compared with cooling or the use of pesticides.

Yet another means of protecting stored products is to keep them in airtight underground pits or sealed metal or concrete silos, or in collapsible butyl rubber silos. This is effective because in dry grain (<14 per cent moisture content) any insects deplete the oxygen to less than 2 per cent and die. Airtight stores may be purged with carbon dioxide or nitrogen to accelerate the process.

Biological control — the use of parasites, predators, or pathogens — is normally not practicable with storage insects and mites; the reason is that the level of population at which parasites or predators effect control is too high, and success is uncertain compared with chemical methods. Predatory mites have, though, been used experimentally in combination with fumigation, as a means of controlling mite pests of grain in Czechoslovakia.

### Current problems

At present, the chief means of reducing losses from storage pests are improved methods of storage and the wider use of chemicals. The use of chemicals is, however, being constrained in two ways. First, there is the development of resistance to contact insecticides and to a lesser extent, to fumigants of many species of storage pests (see Chapter 16). Such resistance has occurred in many parts of the world. In Britain no serious malathion-resistant saw-toothed grain beetle infestations have been found so far; but once this happens, the only effective alternative known at present is pirimiphos methyl — and eventually resistance to that material may well develop. This raises another point: the cost of development and safety testing of a new pesticide is so great that it is doubtful whether it is justified solely for the relatively small market for controlling pests in storage, unless the chemical has also a wider use for controlling field crop pests or those of public health.

A second problem is the tendency, both nationally and internationally, to restrict the use of pesticides, particularly on the initiative of developed countries. Such restrictions are tending to be applied even in developing countries, where the risks from ingestion of small amounts of insecticide such as malathion, DDT, or lindane, used to protect grain stocks, are much less than the risk of undernourishment or famine resulting from destruction of food by pests.

## References and further reading

1  BURGES, H.D. and BURRELL, N.J. Cooling bulk grain in the British climate to control storage insects and to improve keeping quality. *J. Sci. Fd Agric.* **1**, 32-50 (1964).

2  CHRISTENSEN, E.M. (ed.) *Storage of cereal grains and their products,* 2nd edn, A.A.O.C. (St Paul, Minnesota (1974).

3  FREEMAN, J.A. Problems of infestation by insects and mites of cereals stored in Western Europe. *Ann. Technol. agric.* **22**, (3), 509-30 (1973).

4  —, Infestation of food in temperate countries with special reference to Great Britain. *Outl. Agric.* **8**, (1) 34-41 (1974).

5  HALL, D.W. Handling and storage of food grains in tropical and sub-tropical areas. *Agric. Dev. Pap. FAO.* **90**, (1970).

6  MONRO, H.A.U. Manual of fumigation for insect control. 2nd edn. *FAO. Agric. Stud.* **79**, (1969).

7  MONRO, J.W. *Pests of stored products.* Hutchinson, London (1966).

8  WILKIN, J.R. and HOPE, J.A. Evaluation of pesticides against stored product mites. *J. stored Prod. Res.* **8**, (4), 323-7 (1973).

# PART III

## COMPLICATIONS AND ALTERNATIVES

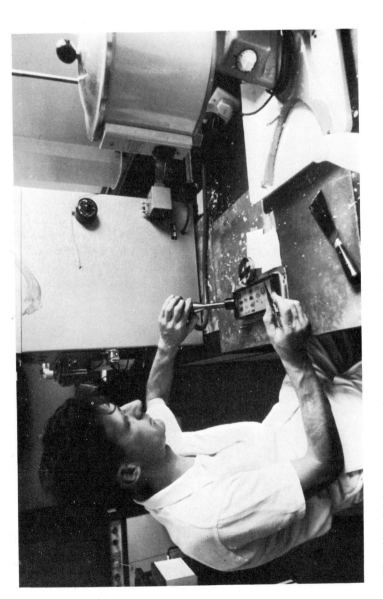

Plate 15.1. A bioassay being carried out on animal tissues. Before a new pesticide is put on the market, laboratory tests on animals give a reliable picture of its likely toxicity to man. (Shell photograph)

# 15 Hazards to people

by J.M. Barnes

Compared with other industries, agriculture has a high accident rate in the developed countries. As well as the more obvious severe and fatal accidents with tractors and other heavy machinery, there are hazards from chronic trauma to the spinal column from ill-sprung tractors driven over rough ground and damage to hearing from noise. Farm animals may inflict physical injury as well as transmit infections such as brucellosis or anthrax, while a serious and unpleasant infection, leptospirosis, may be acquired from wild rats.[7]

Acute or subacute injury to the lung may occur from exposure to the oxides of nitrogen emitted from silo towers, and the farmer and members of his family may also develop a chronic, recurrent, and ultimately fatal pulmonary mycosis from the inhalation of spores from mouldy hay: 'farmers lung'. The accident rate among children on farms is unnecessarily high because they are allowed, and sometimes even encouraged, to play with farm equipment as they never could with other industrial machinery.

Thus the arrival some 30 years ago of a few acutely poisonous pesticides among the agricultural community represented an addition to a wide range of dangers which continue to receive much less attention than they deserve. There was no reason to expect that procedures recommended for the safe handling of pesticides would be followed any more conscientiously than those for the safe handling of other sources of hazard on the farm. The remarkable record of the safe use of pesticides on British farms therefore bears witness to the very limited toxic hazard that pesticides present.[3, 12] In some of the developing countries the use of pesticides may involve the first contact of a community with any toxic chemical and some of the consequences flowing from this are referred to below. Regrettably few communities make any attempt to record accidents from pesticides among agricultural workers.[9]

## Nature of the toxic effects of pesticides

The immediate toxic effect of a pesticide in a poisoned person is usually related to its mode of action on the target organ. Nicotine,

lindane, dieldrin, endrin, parathion, and aldicarb produce by different
biochemical mechanisms an acute disturbance of the nervous system
which, if it is severe enough to interfere with respiration, may be fatal.
The symptoms and course of poisoning differ, but if the patient
recovers, function is fully restored and no permanent injury results. The
absence of any structural damage to the nervous system or elsewhere is
confirmed by observations on experimental animals exposed to larger
or more frequently repeated doses than are encountered by the
occasionally poisoned person.

Pesticides of another group interfere with energy transformation in
the tissues. There is a large group of substances interfering with
oxidative phosphorylation and of these only dinitro *ortho*-cresol
(DNOC) and pentachlorophenol are known to have killed people. Again,
there is no irreversible damage remaining in people or animals who
recover. Of the poisonous pesticides still in use, methyl and ethyl
mercury fungicides are outstanding in producing permanent injury to the
nervous system, while the fumigants carbon tetrachloride and methyl
bromide may in cases of severe poisoning produce some damage to the
kidney and nervous system respectively. With these exceptions, those
pesticides that are toxic to mammals behave as acute poisons depending
for their effects upon a disturbance of function and not the destruction
of vital tissues.

The great majority of pesticides in widespread use are not, however,
serious poisons for people. They are selectively poisonous for the
target organism. Sometimes the basis of their selective toxicity is well
understood. In the case of malathion the mammal but not the insect has
a metabolic pathway which hydrolyses malathion before it is activated
to the toxic maloxon. In the case of many herbicides the site of action
is a system that exists in plants but has no counterpart in animals, and
animals degrade and excrete the herbicide by processes that exist to rid
the organism of the dozens of unwanted substances that are absorbed
from a variety of sources during normal life.

Occasionally a substance which is normally used quite safely may
show an unexpected effect when absorbed in a bigger dose. Paraquat is
an example of a substance that has been safely applied over millions of
hectares by thousands of people without harm to them other than
occasional injury to the eye or skin if unprotected from splashes of
concentrate. However, less than a mouthful of the concentrate accidentally swallowed may cause a severe, progressive, fatal injury to the lungs.
This toxic effect was observed in the animal tests carried out before

paraquat was marketed. The point to be emphasized here is that the 'toxicity' – the capacity to produce damage – is one thing and the 'hazard' – the probability that damage will be produced under any particular circumstances – must be considered separately. The introduction of any new chemical into the human environment must be phased and this is the purpose of most pesticide regulatory schemes. Substances cannot be simply labelled 'safe' or 'dangerous'. Their toxic properties together with their manner of use will determine whether they can be used safely.

The evidence available from the data compiled for over 20 years in the U.K. is that pesticides can be used safely and that the deaths from overturning tractors alone are greater in any one year than all reports of illness arising from exposure to pesticides. There have been no deaths from pesticide poisoning. Reports are compiled of all cases where men have been off work for illness attributed to exposure to pesticides. The majority of these cases arise from irritation or inflammation of the skin or eyes resulting from splashes while mixing concentrates. Among the substances used as pesticides there are a few which can induce sensitization in people who come in contact with them; such individuals may have to avoid all future contact with a particular compound.

There are a number of relatively common serious human diseases for which no general cause has yet been established: leukaemia, aplastic anaemia, and peripheral neuritis are three examples. When a patient presents such an illness the doctor often seeks by enquiry some possible cause. As a result there are scattered in the medical literature claims that exposure to a pesticide has led to the development of one or other of these diseases. There is no epidemiological evidence to support such claims; in other words, the disease is not seen among those most heavily exposed or surviving acute poisoning by the compound. The aplastic anaemia reported in individuals said to have been exposed to BHC has been attributed to benzene which does cause aplastic anaemia; but BHC is based on cyclohexane which does *not* cause anaemia. There is no epidemiological evidence that exposure to pesticides causes cancer.

## Circumstances under which poisoning by pesticides has taken place

Although large amounts of a pesticide in its most concentrated form are encountered in the manufacturer's plant, it is not difficult to protect workers from exposure in a modern chemical factory, and few people may be at any risk when a process is largely enclosed. Protection may be less adequate in some formulation plants, in which the active pesticide is mixed with inert dusts or made into a solution or emulsifiable concentrates, particularly when dry powders and dusts are being produced. Acute poisoning has been reported among men formulating some of the more toxic insecticides.

There have been reports from all parts of the world of poisoning among those applying pesticides in the field.[10] The exposure is greatest among those who dilute and mix the concentrates and the same applies to those who load aircraft for aerial applications. Occasionally pilots have been poisoned from leaks from pipes in badly designed spraying aircraft. Those who maintain spraying machinery may also be at risk if the machines are not properly cleaned at the end of the operations. Knapsack sprays and pumps may leak and contaminate the clothing of those who operate them. There have been occasional reports of illness among agricultural workers who have been pruning, thinning, or picking recently sprayed crops such as peaches and grapes. Such episodes have been confined to crops sprayed with parathion, one of the most toxic insecticides still in general use.

While fatal and non-fatal poisoning has followed the careless application of pesticides for the proper purpose of helping the better growth of crops or the storage of harvested products, there are other circumstances under which casualties, often on a dramatic scale, have occurred. Seed that had been dressed with pesticides to ensure better germination and growth, particularly if it is not issued to the farmers for sowing until after the correct sowing time, has been eaten in the place of their own grain which they had already sown. In Iraq in 1972, hundreds of deaths occurred and thousands of men, women, and children were permanently disabled after consuming grain dressed with a methyl mercury fungicide.[1]

In other episodes dressed seed has been stolen and sold as edible material. In one instance the dressed seed was fed to livestock and a family who ate a pig became severely affected. It is not surprising that in a recent report WHO and FAO recommended the withdrawal of all methyl mercury seed dressings except for use in seed 'banks'.[8]

Other outbreaks of poisoning have resulted from the contamination of human food, usually flour or sugar in hessian sacks, from leaking or broken containers in ships and vehicles. Even the separation of the pesticide from the food by storage in different holds (which, however, by no means always took place) was not enough to prevent one out-break where the pesticide concentrate dripped through the floor of the hold and soaked into sacks of flour.[13]

There have been many cases of poisoning from pesticides among children, though many fewer in the U.K. than in the U.S.A.[6] Whether or not this difference is linked to the greater freedom or enterprise of American children is not certain. What is certain is that many such episodes stem from the gross carelessness of adults in leaving so-called 'empty' containers available as playthings.

Insecticides found to be obviously so effective against an agricultural pest have been incorrectly and dangerously applied for the eradication of others, such as bed bugs or body lice, with a consequent dangerous or fatal poisoning of members of a household. In some communities an 'empty' 25-litre drum may represent a valuable addition to household equipment; if the first use to which such a drum is put is the mixing of drinks for a wedding party, the outcome will be disastrous.

Finally, the presence of poisonous pesticides in a community may provide a new and convenient means of suicide or homicide.

Thus, it is clear that pesticides can and do kill or seriously poison people. The most dramatic and serious episodes have resulted from misuse, whereas their legitimate application on an ever-increasing scale for pest control purposes appears to be associated with fewer reports of accidents. It is worth turning aside to enquire into the reasons for this.

**Developments which have reduced hazards**

There is little doubt that knowledge about the ways in which people are actually exposed when they apply pesticides has been crucial in the development of safer methods. Because so many applications of pesticides are by means of sprays, either of suspensions of wettable powders or of diluted emulsions, it has been widely believed that pesticides are dangerous because they might be inhaled. Careful studies with measurements made during various operational procedures showed conclusively that the main route of exposure was the skin and not the respiratory tract.[14] In operations that varied from indoor spraying to applications in orchards, ground crops, and the soil itself the amount of material on the skin or clothing exceeded many-fold the quantity that would have entered the mouth and nose. The explanation is simple: namely, that particles that will settle on leaves or any other surface must exceed a certain mass, whereas only very much smaller particles which would not impinge and settle on surfaces can enter the lungs. Thus, the only pesticides that are a potential danger by inhalation are those few preparations which are released as smokes or vapours and against which special protection by respirators can be provided.

Knowledge that the skin is the principal route of exposure for pesticides means that attention can be focused on simple protection of the exposed areas and an insistence on regular washing during any prolonged period of exposure. Except when handling concentrates, heavy-duty rubber protective clothing is not needed − and would be intolerable to wear in many circumstances. However, light cotton will prevent droplets reaching the skin. Another consequence of this knowledge has been the attention paid during the toxicological testing of new pesticides to the ability of a substance to pass through the intact skin. This can rarely be predicted from chemical structure or physical properties of a compound. Those pesticides that have proved dangerous to use (other than volatile ones) have an acute toxicity by the dermal route that is not greatly different from that by the oral route, whereas for the great majority of compounds in common use it may be difficult to elicit any serious acute toxic effects in animals after skin application alone.[4] In accepting new compounds for use as pesticides attention is paid to their acute dermal toxicity to laboratory animals. In some countries it is feasible to insist that certain protective measures are provided for those employed in the application of the more toxic pesticides. Where compulsory protective measures, including the provision of special garments, need not be recommended when less toxic analogues are applied, an important sales incentive for the development of less toxic pesticides is provided. However, where very toxic substances have an important role as pesticides, attention has been paid to safer formulation such as adsorption on to inert granules or incorporation into other solid bases to provide a slow-release device. The fact that fewer safety precautions are needed for the use of such preparations may partly override their greater cost.

Another important factor in reducing hazards is the greater control being exercised by a number of governments over the introduction, marketing, distribution, and labelling of pesticides, accompanied by better field services and advice to users. It is becoming less frequent for the local salesman to be the only source of information about a new pesticide. Where there is a more intensive and commercially important agricultural enterprise in a region there may be opportunities for training in the practice of using pesticides. This may often be done to make certain that the correct applications are made to ensure that the residues of pesticide in the treated crops do not exceed the limits set by prospective buyers. All such training is likely to encourage the use of methods providing greater safety for those applying the materials.

Knowledge of the mode of action of the anticholinesterase insecticides* has made it possible in the case of the organophosphorus compounds to devise a simple field method for detecting exposure long before serious or dangerous toxic effects appear, and thus permit a man to be withdrawn from further danger. Such arrangements are practicable only during the trial introduction of a compound. There are, however, effective antidotes to acute poisoning by both the anticholinesterase compounds and the convulsive chlorinated hydrocarbon insecticides, so that if agricultural workers do become ill and are taken to hospital it is unusual for them to die. However, the best protection against poisoning by new pesticides is the large amount of information that is acquired about their toxicity before they ever become commercially available.

**Value of laboratory tests on animals.**

During the past 50 years it has become increasingly customary to examine the toxicity of new substances on laboratory animals before the compounds are introduced into the human environment. Although such screening was first done, albeit on a very limited scale, on compounds destined for use in industrial processes where men might be exposed to vapour or liquids, there were until recently many compounds in wide commercial use for which there were few toxicity data. For modern pesticides there has always been some information on mammalian toxicity. This is exemplified by the early studies of the effects of DDT on the common laboratory species including monkeys, as well as on man himself. These investigations were carried out within two or three years of the introduction of DDT in 1945.[5] The early organophosphorus anticholinesterases were recognized as close chemical relations of lethal 'nerve gases' and no one was surprised at the great toxicity of parathion and TEPP. The early casualties among people who first applied parathion, often with no more regard to safety than they gave to the use of DDT, encouraged the search for compounds that killed insects but were less acutely toxic to mammals. The result is that now there are compounds such as malathion or phoxim for which the margin of safety is enormous. This is related to differing rates of activation and degradation in the whole animals − insects or rats. Such vital information would never come to light if preparations such as tissue culture replaced rats in tests for toxicity.

One of the earliest herbicides, DNOC, caused some deaths from

*See p. 263.

acute poisoning about 1950 among those spraying cereals in hot weather. When used as a slimming agent 20 years earlier, DNOC killed some patients but it was not until it was used as a selective herbicide that a proper study of its mammalian toxicity was undertaken. By contrast, 2,4-dichlorophenoxyacetic acid was shown in the same period to have a very low mammalian toxicity, and it has had a long history of safe use. From this early work there developed an increasingly complex pattern of tests on laboratory animals exposed both acutely and by repeated or prolonged application; and it can now justifiably be stated that laboratory tests on mammals give a reliable picture of the likely toxicity to man, although animal tests may not indicate that a compound can induce skin sensitization in man.

Toxicity tests on animals have thus made it possible to foresee and forestall the hazards of acute poisoning from occupational exposure; but this has not prevented people from expressing anxiety about the possible late effects of a prolonged exposure to small amounts of a pesticide that occasionally remain as a residue in food prepared from treated crops. Since tests on animals give a good indication of the possibility of acute poisoning it is irrational to assume that they are inadequate as indicators of possible chronic toxic effects. When pressed for examples of a toxic effect from a chronic exposure to low doses most people mention only cancer.

However, it is also known from extensive studies on chemicals other than pesticides that laboratory animals can respond by producing cancers resembling those seen in man. The response of animals to a prolonged heavy exposure to a new substance thus provides a reasonable basis for deciding whether the compound has the properties of a chemical carcinogen. At the present time, however, a great deal of disagreement exists about the significance of lesions produced in the livers of mice by a whole range of substances including DDT. While the tumours in the livers of these mice have many pathological features of liver cancers, the behaviour of the lesions and of the animals as a whole do not resemble those accompanying the evolution of a true cancer. Any compound being developed as a pesticide which during the course of toxicity tests on animals showed the capacity to produce cancer would be withdrawn from further development. At the present time there are no grounds for labelling as carcinogens DDT and half a dozen other pesticides that in large doses produce liver tumours only in mice.

Part of the alarm about chemical carcinogens as hazards rests upon the undoubted fact that for people exposed to carcinogens in industry, just as for many smokers, the first evidence of a toxic effect is the appearance of the cancer, possibly at an incurable stage. Coupled with this fear is a widespread belief that there is no 'safe dose' of a chemical carcinogen because a single molecule homing on a sensitive DNA molecule might start a cancer. Recent work on the capacity of damaged DNA to undergo repair indicates, however, that there is likely to be a dose-response relationship for chemical carcinogens similar to that for other toxic substances. With respect to other toxic effects, the very extensive laboratory tests, including lifetime studies on rats continuously exposed to pesticides now in use, amply demonstrate the existence of fairly steep dose-response relationships. This indicates that if the people receiving the heaviest exposure during the application of a pesticide show no ill-effects, then it is extremely improbable that those absorbing minute traces in food will suffer any ill-effects.

## Contemporary evidence for assessing hazards

There are unfortunately no good figures for the incidence of acute or any other sort of poisoning of people by pesticides, but from poison control centres it is evident that pesticides are responsible for only a very small proportion of cases of poisoning, both fatal and non-fatal, from chemicals.

From hospital reports there is no evidence that unusual or prolonged effects follow recovery from the acute pesticide poisoning from whatever cause, although the follow-up of acute poisoning is only for very short periods.

The recognition of long-term toxic effects from substances to which people are occupationally exposed is difficult, even when it is a rare tumour in a comparatively closed population. It is therefore not feasible to set up a study that is likely to provide evidence, negative or positive, that exposure to a pesticide can eventually lead to cancer in man. Agricultural populations are widely scattered, and hospital and general medical services are likely to be minimal in rural areas. Most important of all, this exposure to pesticides will be mixed, intermittent, and to a constantly changing group of compounds. While this last factor makes a follow-up difficult, it also protects the individual, for he will not be exposed to any particular substance over a long term. The only groups of people who have had a continuous and unusually heavy exposure to a single pesticide are the men who have formulated DDT since 1948 and others in malaria control programmes who have applied

DDT as an indoor spray year after year. One group of formulators has been followed-up in the U.S.A., and WHO are attempting to do the same for groups of spraymen in Brazil and India — but establishing the cause of death in a country where autopsies are not done has its problems.

As mentioned above, there has been much concern expressed about pesticide residues in food.[2] This stems partly from the widespread practice of having 'tolerances' or limits which pesticide residues in treated crops should not exceed. It is popularly believed that if these 'tolerances' are exceeded then that food is dangerous to eat. In practice a tolerance is established as the level which need not be exceeded if the crop has been treated as recommended in order to control a particular pest. With the improvement of analytical techniques the practice of doing analyses on market basket samples has increased. In this procedure the food as actually consumed is analysed. When this is done it is found that most pesticide residues have disappeared: for example, 'Extensive data show that most pesticides leave little or no residue in the majority of foods and most of the residues present are removed in storage, preparation and processing.'[11]

## Outlook for the future

While the continued clamour about hazards from pesticides may well seem to be quite out of proportion to the dangers that accompany their use, there is no doubt that attention will remain focused on better and safer methods of using them. Unfortunately it is unlikely that such advances will take place first in those communities where the dangers are greatest. Where there is little or no control of the introduction and distribution of pesticides the cheapest and most effective materials may be used by people untrained and ill-equipped to apply toxic compounds, but it is to be hoped that eventually most of the really dangerous compounds will no longer be marketed by reputable organizations.

The more extensive use of less dangerous materials should not be the pretext for abandoning care in the handling and application of pesticides. They should all be treated with respect, because it is never possible to be absolutely sure that substances shown to be safe in the short term will not have some deleterious effects in the long term if exposure has been heavy.

It is vain to hope for the ultimate replacement of existing pesticides by compounds that are more highly selective in their action, although for many herbicides this is already the case. What we may reasonably expect is that the most persistent materials will be replaced by more biodegradable counterparts. While the control of pests continues to depend extensively on the use of chemical agents, it is reasonable to predict that the emergence of resistant strains of pests will ensure repeated changes in the nature of the substances used, thus avoiding long-term exposure to any particular pesticide.

*     *     *

The toxic effects of all pesticides continue to be scrutinized with great care, and as much is known about their behaviour and metabolism in animals as is known about many things to which people are more widely and more heavily exposed.

There have been cases of acute poisoning under circumstances which occasioned no surprise, either because of the toxicity of the particular pesticide involved or the gross carelessness and disregard for elementary safety associated with the incidents. Proper methods of handling avoid serious risks to people, and the probability of accidents diminishes with a greater general awareness of potential hazards.

The effective use of pesticides does not lead to the gross contamination of human or domestic animal food. The minute traces measured in parts per million that are sometimes found in food are of no toxicological significance even in the case of the poorly biodegradable organochlorine compounds. The use of such persistent compounds will gradually disappear for general aesthetic reasons and no case based on the toxicological hazards they present can be made for any immediate or drastic action to effect this change.

### References

1  BAKIR, F. *et al.* Methyl mercury poisoning in Iraq. *Science, N.Y.* **181**, 230-41 (1973).
2  BARNES, J.M. Pesticide residues as hazards. *PANS.* **15**, 2-8, (1969).
3  ── Toxicology of agricultural chemicals. *Outl. Agric.* **7**, 97–101 (1973).
4  GAINES, T.R. Acute toxicity of pesticides. *Toxicol. Appl. Pharmac.* **14**, 515-34 (1969).
5  HAYES, W.J. Pharmacology and toxicology of DDT. In *DDT Insecticides.* Ed. Mueller, P. Vol. 2, pp. 11-247. Birkhauser Verlag, Basel (1959).
6  ──, and PIRKLE, C.I. Mortality from pesticides in 1961. *Archs. Environ. Hlth.* **12**, 43-55 (1966).
7  Health hazards in agriculture. Proceedings of 23rd Conference of the British Occupational Hygiene Society. *Ann. occup. Hlth,* **12**, 63-138, (1969).
8  Mercury and alternative compounds as seed dressings. Report of a Joint FAO/WHO Meeting. *Tech. Rep. Ser. Wld. Hlth. Org.* **555**, (1974).
9  *Modern trends in the prevention of pesticide intoxications.* Report of a Conference, Kiev 1971. World Health Organization, Regional Office for Europe, Copenhagen (1972).

10 *Pesticide Abstracts* (formerly *Health Aspects of Pesticides*). Published monthly. U.S. Environmental Protection Agency, Washington, D.C.
11 *Pesticide residues in food.* Report of the 1970 Joint FAO/WHO Meeting. *Tech. Rep. Ser. Wld. Hlth. Org.* **474,** (1971).
12 *Safety, health, welfare and wages in agriculture. Annual Reports of Ministry of Agriculture, Fisheries and Food.* H.M. Stationery Office, London.
13 WEEKES, D.E. Endrin food poisoning. *Bull. Wld. Hlth. Org.* **37,** 490-514 (1967).
14 WOLFE, H.R., DURHAM, W.F., and ARMSTRONG, J.F. Exposure of workers to pesticides. *Archs. Environ. Hlth.* **14,** 622-33 (1967).

# 16  Pest resistance to pesticides

by James R. Busvine

I can clearly remember first hearing of DDT-resistant houseflies from the late Professor Missiroli in Rome in 1947. This was only five years after our initial tests of the new insecticide and, such was the reputation of DDT at that time, it was difficult to accept his story; but he kindly supplied me with some of the portentous flies and we soon verified the fact. Furthermore, it was not long before reports began to multiply of similar immune strains of flies and other insects. The obvious seriousness of the matter began to change the common usage of the word 'resistance' among those concerned with pest control. Formerly, we had used it to describe normal variations in tolerance due to environmental changes, to which insects are very prone; alternatively, it was used as a characteristic of particular species of pest, some of which can withstand much higher doses of poison than others. Now, however, it is convenient to use the words 'susceptibility' or 'tolerance' for these familiar concepts and to reserve the word 'resistance' for abnormal strains selected by extensive pesticide usage.

The phenomenon is not entirely novel. As long ago as 1911 there was evidence that extensive fumigation of citrus trees had selected a strain of scale insect resistant to hydrogen cyanide gas; and there were perhaps four or five other cases before the Second World War. But the growth in incidence of resistance since 1947 was very much larger and faster, as a result of the introduction on a vast scale of new synthetic insecticides. This growth of resistance is illustrated in Fig. 16.1 by the numbers of species involved; it refers to insects of medical importance only, because their early resistance was better documented than that of the agricultural pests. It seems that 52 species of agricultural pests had developed resistant strains by 1958 and 228 by 1969: the first figure about equalled that of medical pests; the second far outstripped them.

Within a few years, resistant houseflies had been reported from very many countries; and similar troubles had been reported for lice, fleas, certain culicine mosquitoes, and various agricultural pests. Without discussing the practical importance of these figures at this point, it need hardly be said that the phenomenon soon caused concern. In particular, officials of the World Health Organization, who

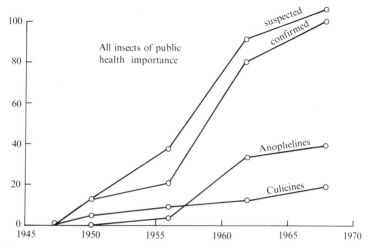

**Fig. 16.1.** Numbers of species of arthropods of public health importance which have developed resistance to insecticides.

were concerned with insect-borne diseases, became justifiably anxious at failures in the cheap and effective new weapons being so widely used to control disease vectors. Accordingly, members of the Expert Committee on Insecticides were asked to advise on the problem in 1956 and at subsequent meetings in 1957, 1959, 1962, 1969, and 1975. Somewhat later, the Food and Agriculture Organisation became concerned about resistance in pests of agricultural and veterinary importance and arranged similar meetings annually from 1965 onwards.

The first task of these international organizations was to assess the situation on a global basis. This, however, proved very difficult in the earlier discussions, because the reports of resistance received were so very varied, ranging from mere subjective impressions to cases carefully authenticated by laboratory experiments. There was clearly a need for reliable test methods to detect genuine resistance; and since many of the pests concerned were widely distributed, it was essential that the methods should be internationally standardized. In this way, the results obtained by a worker in one country would be meaningful and comparable with those in another. Accordingly, both WHO and FAO have, through the years, developed and published standardized methods for the more important pests of medical, agricultural, and veterinary importance. The actual manipulations in the tests are kept as simple as possible, so that they can be conducted without elaborate apparatus.

194

In many cases, insects are simply confined with papers impregnated
with various insecticides at predetermined rates for specific periods.
The WHO actually provided reputable field workers with such
impregnated papers in complete test kits and has received many
thousands of sets of data (necessitating analysis by computer) so that
up-to-date information on resistance throughout the world has been
available for insect disease vectors.

Both WHO and FAO have tried to foster research on the fundamental
aspects of resistance, which had in any case attracted the interest of
many individual scientists. As long ago as the early 1950s, there had
been enough scientific investigation of resistant strains for the general
nature of the phenomenon to be understood. The idea of resistance
being acquired by insects as a result of sub-lethal doses (by analogy
with certain diseases) was soon abandoned. The character was
demonstrably hereditable and already in 1951 my former colleague
Dr. Mary Harrison had shown that it segregated according to normal
Mendelian laws. It was evident that the resistant strains emerged
because of the selective survival, after the wide use of insecticides, of
certain individuals which happened to possess genes giving them more
or less immunity. On this basis, it was possible to predict the conditions
conducive to the development of resistance; and despite much more
intensive research in intervening years, three factors which I proposed
in 1954 seem still relevant. They were:

(1) the frequency at which resistant genes happen to occur in the
pristine population, and their effectiveness;

(2) the intensity of selection, i.e., the magnitude of the population
exposed to insecticide and the proportion killed; and

(3) the number of generations per year.

Experience has shown that factor (2) is overwhelmingly important.
Resistant strains have developed mainly in important pests which have
been constantly attacked by insecticides over wide areas. Certain
aspects of this are reflected in Fig. 16.1. The sharp rise in cases of
resistance during the period 1956-62 was partly due to wide use of
dieldrin (to which resistance develops quickly) in the WHO programme
of malaria eradication during that period. The fact that resistance is still
somewhat localized even in such cases suggests that resistance genes are
not universally distributed; where they do not occur, insecticides retain
their effectiveness. Few people, however, have had the time and patience
to search for rare resistance genes in untreated populations. The
'effectiveness' of resistance genes mentioned in factor (1) refers not

only to the degree of protection afforded but also to their dominant or recessive status. This undoubtedly affects the rate of selection, because recessive genes confer no advantage in the heterozygous state (and homozygotes are very rare when the gene is uncommon); whereas dominant genes respond to selection even in the heretozygous condition and rise to prominence more quickly.

Another early concept, which has remained largely unchallenged, is that of the specific nature of resistance, due to the possession of a particular physiological mechanism, rather than the concept of a tough 'super insect'. Two important and highly specific defence mechanisms which soon became prominent were (i) an enzyme system which detoxified DDT (and a few similar compounds) by breaking it down to a non-toxic compound, 'DDE'; and (ii) a change in an unknown vital system which made insects insensitive to BHC and to dieldrin-type insecticides. Each system is effective in protecting an insect against a group of related poisons and any member of such a group will tend to select for that particular kind of resistance. This is the basis of 'cross-resistance'; that is, the automatic involvement of other insecticides when resistance has developed to one member of a related group.

Since the two original types of resistance were so specific, it was possible to maintain effective control by changing to an insecticide in the uninvolved group. Eventually, however, a number of species were able to develop 'double resistance' to both groups of chlorinated insecticide. It was then still possible to change to one of the organophosphorus or carbamate insecticides (or, indeed, to new synthetic pyrethrum-like compounds which were developed later). However, many species have shown themselves capable of developing additional resistance mechanisms to protect themselves against these alternative insecticides, thus producing triple or quadruple resistant strains.

### The nature of resistance

Not unnautrally, the progress of scientific techniques during the past two decades has greatly widened and deepened our understanding of the nature of resistance. Techniques using radioactive isotopes or thin-layer or gas-liquid chromatography have been used to determine in detail the chemical degradation pathways in enzymatic detoxication of poisons in insect tissues. Some of the enzyme systems responsible have been isolated and characterized. These biochemical researches have revealed a considerable number of detoxicating systems, in addition to the well-known DDT-DDE mechanism. Most of these new discoveries relate to organophosphorus and carbamate insecticides, which undergo varied and complex chemical changes in insect (or vertebrate) tissues. Genetically based changes in an enzyme system involved in such a process can alter the balance of poisoning and detoxication. Most forms of resistance depend on such a change, although in some cases there is evidence of reduced sensitivity at the site of action (similar to BHC-dieldrin resistance). In addition, changes in insect cuticle have been found, which tend to reduce the penetration of new insecticides, nearly all of which can act by contact after penetration.

While these advances in toxicology were proceeding, there were refinements in genetical techniques. In the early days, we merely crossed normal and resistant strains and observed the segregation of the resistance factor in the progeny (noting, of course, the presence of dominance or recessivity). Modern studies employ susceptible colonies with recessive marker genes (affecting eye colour, etc.) on various chromosomes, so that when these are crossed with resistant strains it is possible to assign different resistance genes to specific linkage groups. By appropriate genetic manipulations, resistance mechanisms dependent on different genes can be separated or combined.

The amount of research on various aspects of insecticide resistance published during the past 25 years would probably fill a volume of an encyclopaedia. Many of the investigations have been illuminating and technically impressive and it is sad to conclude that no satisfactory simple way of overcoming resistance has been found. One truly scientific solution, whose early hopes have not been fulfilled, depended on the fact that several forms of resistance depend on enzymatic detoxication systems, which can be blocked by appropriate chemicals. For example, the compound dicofol inhibits degradation of DDT to DDE; and when mixed with DDT it was able to restore its potency against some early resistant strains. Unfortunately, insect genetic resources are so adaptable that alternative resistance systems are selected if one is blocked; and it becomes impossible to cope with them all.

Experience has shown that once resistance has fully developed to a particular pesticide it has never yet been possible to reverse the process, either by ceasing to use the pesticide or by any other means. One is tempted to remember the physicians who attended Henry King ('They answered as they took their fees, "There is no cure for this disease" '). Yet much has been learnt which enables us to understand resistance more intelligently and to concentrate on the factors which promote or delay it. All the evidence shows that resistance will not develop without intensive and widespread use of a pesticide; and sometimes pesticide usage is excessive and wasteful. There are several instances of resistance in insect vectors (such as mosquitoes) developing as a result of heavy agricultural uses (for example cotton crops, sprayed many times each season). The most sensible course is to limit usage to the minimum quantities essential to prevent disease transmission or really serious crop losses. This suggests that we should take a much closer look at the actual impact of resistance in terms of public health and economics. Such assessments are difficult and can be attempted only by informed experts directing enquiries to competent field operators in many countries usually through international agencies. I have myself been involved in two such surveys on behalf of WHO; while two American professors (H.T. Reynolds and G.P. Georghiou) were in 1975 concluding one on behalf of FAO.

## The impact of resistance on control of disease vectors

Two surveys have been made (in 1968 and 1975) questioning public health authorities (about 100 and 200 respectively), mainly in the tropics. They showed how extensively the control of disease vectors has depended on the use of organochlorine insecticides (especially DDT, BHC, and dieldrin) in the past two decades. However, in some countries, areas of double resistance have invalidated both organochlorine groups; and this has severely impeded the control of many insect species and of the diseases they control. They involve nearly a score of anopheline vectors of malaria, the urban mosquito vector of yellow fever *(Aedes aegypti),* the common house mosquito which spreads filariasis *(Culex fatigans),* the louse vectors of typhus, and the tropical rat flea which carries plague. Housefly resistance occurs in all parts of the world. Diseases of the bowel and the eyes which flies spread are, however, transmitted by other means as well, whereas diseases such as malaria and typhus have no alternative vectors.

Malaria is outstanding in severity and magnitude, since some 2000 million people live in potentially malarious areas. Vigorous efforts in the past 25 years have reduced the annual incidence from 300 million (with 3 million deaths) to about 50 million (with 1 million deaths). This was largely due to residual insecticides, especially DDT; but the success was mainly in peripheral zones, leaving many huge, hyperendemic areas untouched. Further progress is very uncertain, largely because resources are inadequate but also because of resistance and, it must be said, the partial abandonment of DDT owing to the adverse propaganda of environmentalists.

Certain other health problems (onchocerciasis, Chagas' disease, trypanosomiasis, leishmaniasis) have not so far been gravely affected by resistance; but it is difficult to be sure of the continued reliability of the organochlorines, so cheap, safe, and otherwise efficient. Indeed, some of the lack of resistance is due to the meagre scale of past control measures; if these are substantially expanded, resistance is likely to develop.

## Impact of resistance on agricultural and veterinary pests

A global survey was organized in 1965 by an FAO Working Party on Resistance. Some attempts were made to keep information up to date, but by 1973 it was evident that another complete survey had to be done. By late 1974, over 300 replies had been received, about 20 per cent of them concerning new examples of resistance. Not only had the severity of the resistance grown and more countries become involved, but the trouble had spread to different groups of insecticide. At the time of the earlier survey, very few pests had become resistant to organophosphorus or carbamate insecticides; but these are frequently mentioned in the more recent reports, several of which concern pests resistant to all the commonly used modern groups of pesticide, rendering control extremely difficult.

It seems likely that this situation has been partly brought about by excessive dependence on pesticides, because of their convenience and efficiency, at the expense of other control measures such as crop rotation, the use of immune varieties, and changes in agricultural practices. As a result, the growth of resistance in certain regions has reached a point where control of the principal crops can no longer be • assured by pesticides alone. Cotton pests are resistant in several areas (e.g., in the southern U.S.A., the bollworm, the bollweevil and the pink bollworm; in the Near East, mainly the cotton leafworm). Several rice pests have developed multiple resistance, especially in Japan; notably the rice stem borer and two leafhoppers. In West Africa, capsid resistance threatens the all-important cocoa crop; in the West Indies, resistant sugar cane froghoppers are a menace. In Australia, the beef industry is hampered by the near-immunity of cattle ticks, especially in Queensland, while sheep farming is plagued by resistant sheep blowflies.

Resistance problems of other pests are locally severe in many countries. Vegetable crops are threatened by resistant strains of various species of root maggot flies, by the Colorado beetle, and by the cabbage looper. Among fruit and glasshouse crops, perhaps the main resistance problem consists of various multiple-resistant spider mites in which resistance to any new acaricide develops very fast; and in local areas the trouble affects codling moths and pear psyllids. In North America, cereal crops are affected by resistant cutworms and alfalfa weevils. Tobacco in several countries suffers from resistant hornworms and cutworms. Crops are by no means immune after harvest; a problem of growing severity is resistance in beetle pests of stored cereals, which are being spread round the world in grain shipments.

In the mid-1970s, serious effects of resistance in agricultural and veterinary pests were augmented by several factors, which particularly affected underdeveloped countries. World food shortage required all countries to save crops from the attacks of pests; but current supplies of pesticides became short and more expensive because of market demand and rising costs. Furthermore, environmental propaganda dissuaded many chemical manufacturers from expanding pesticide production. Profits from pesticides are unreliable in the face of the expense of the exhaustive safety testing now required; and the threat of resistance makes the projected economic life of a new pesticide uncertain. In these circumstances, commercial firms are not encouraged to pursue research to discover new pesticides to replace those rendered ineffective by resistance.

## Resistance in harmful organisms other than arthropods

Resistance to control by chemicals has arisen in various pests other than arthropods. Most obvious, perhaps, is the important problem of drug-resistant pathogenic bacteria and protozoa. I shall not, however, attempt to deal with this subject, for the same reason that Dr. Johnson gave for wrongly defining the pastern of a horse ('Ignorance, Madam, pure ignorance'); and also because the basis of heritable selection of such microorganisms is very different from that of other animals and plants. On the other hand, the development of resistance in fungi towards fungicides has some points of similarity with the situation with other pesticides, according to the experts (FAO 1975). I am indebted to similar secondary sources for the following sections.

### *Resistance of fungi to fungicides*

Until about 1970, plant pathologists were little troubled with resistance to fungicides. For many years, heavy metal compounds and some relatively simple organic compounds were used. Such fungicides need to be applied with particular care, since to be effective they have to form a protective film over the whole plant. In contrast, the systemic fungicides developed in the last decade (e.g., oxathiins, pyrimidines, benzimidazoles) penetrate the cuticle and are translocated throughout the plant, so that their action is more efficient, and not merely protective but also partly curative. Because of their more reliable action, they quickly became very widely used.

The older 'protective' fungicides tended to be general cell poisons with low specificity, whereas the newer systemic chemicals generally act on one specific vital function (e.g. benzimidazoles probably inhibit DNA synthesis). Consequently, a relatively simple genetic variant can produce a protective mechanism. As a result of this weakness, and the wide use of modern fungicides, a large number of cases have been recorded in recent years (the FAO Report quoted cites 27 cases published between 1970 and 1973, based on a paper by J. Dekker). Not only is the loss of such efficient pesticides serious, but it has been found that the resistant strains are quite as virulent as the original ones. The few examples of strains resistant to the older fungicides tended to be less aggressive.

### Weeds and herbicides

Considerable use of herbicides is made in modern agricultural practice. There are consequently some examples of the selection of resistant strains of certain weeds; but this particular trouble is for various reasons much less widespread than in other types of pest. Herbicides used to eliminate weeds are usually only broadly specific, tending to remove botanical families or larger groups. What then happens is the invasion of other kinds of weed, rather than the survival of tolerant survivors of the primary species, so that there tends to be a change in the local weed flora.

In the long run, resistant strains of currently sensitive species may presumably be expected to develop; but few herbicides have been extensively used for more than about 15 years. Since many temperate-zone plants have only one generation per year, the numbers of generations selected are not high, especially as seeds may lie dormant for several years. Furthermore, selection is reduced by crop rotation, fallowing, and the use of alternative herbicides.

*Pest resistance to pesticides*

*Nematodes and nematicides*

It is important to distinguish the plant parasitic nematodes from helminthic intestinal parasites of animals. Not only are the chemical control measures quite different, but so is the ecological situation which affects the intensity of selection.

Plant parasitic nematodes are commonly controlled by soil fumigants, usually halogenated hydrocarbons. Despite long usage, no case of resistance has emerged; nor was it developed by 100 generations of experimental selection of one soil nematode. In contrast, there have been about half a dozen examples of resistance in helminths affecting domesticated animals (mainly sheep).

There are possible toxicological and ecological reasons for these different situations. The chemicals used against soil nematodes act rapidly and are unspecific cell poisons, whereas the anti-helminthics are relatively complex organic compounds, with relatively specific action against the parasite and not the host animal, probably with slower penetration; these factors give more opportunity for resistance mechanisms to operate. The strain of *Haemonchus contortus* resistant to benzimidazoles seems to have modified the enzyme system which is the biological target, making it less susceptible.

On the ecological side, it may be noted that the soil fumigants are generally restricted to the soil where plants will be grown. Should there be any selection of local populations of nematodes, the survivors would later interbreed with populations in untreated soil and dilute the selective effect. On the other hand, herds of animals may often be simultaneously treated and restricted to certain areas, so that a considerable proportion of a local population may be selected.

*Rodents and rodenticides*

Two species of rat are serious pests: the Norway rat and the ship rat. Both consume much human food and spoil even more; and the ship rat is a potential reservoir of plague in certain countries. The house mouse is, of course, common but less serious.

For many years, it was the practice to destroy rodents by poison baits containing acutely toxic substances like arsenious oxide, squills, and sodium fluoroacetate, some of them being dangerously poisonous to other animals. To avoid 'bait shyness', it was important to use a troublesome system of pre-baiting and later changing to poisoned baits. About 25 years ago, the slow-acting warfarin type of poison was

introduced. Being simpler to use, safer, and more effective, it soon
replaced the acute poisons.

Warfarin resistance in Norway rats first occurred in Scotland about
1958 and soon afterwards appeared in England. Later, it emerged in
Denmark, Holland and parts of the U.S.A. Warfarin-resistant ship rats
have also appeared more recently (in Liverpool, 1970, and London,
1973). Resistance in house mice has been known since 1960 in the
U.K., Germany, and the U.S.A.

Considering the extensive use of warfarin, the amount of resistance
is not excessive. It is not quite clear why it occurs in some areas but
not others; but, as with insecticides, extensive and persistent use is the
probable cause. About 10 years of selection pressure (20 to 30
generations) appears to be necessary. Although the trouble develops
slowly it is difficult to eliminate when it has become established. A few
new types of poison are now being used against the resistant strains.
One is calciferol (vitamin D) which is highly poisonous when
concentrated.

## Snails and molluscicides

Certain fresh-water snails are the intermediate hosts of the parasite
which causes schistosomiasis in some hot countries. One important way
of controlling the disease is by destroying the snail hosts in ponds,
ditches, and irrigation channels. Molluscicides for this purpose have been
fairly widely used for a decade or two. Copper compounds and relatively
simple organic chemicals (e.g., sodium pentachlorphenate) were used in
the early days, but more complex and more efficient molluscicides are
now being used. Between 1956 and 1961 there were a few reports of
snail resistance in Japan; but the levels were low and true resistance was
never substantiated.

\*　　　　　\*　　　　　\*

In the past decade, we have been reluctantly forced to conclude
that we will have to 'live with' pest resistance to pesticides; and that
will mean a continual struggle. This situation, and the vociferous
complaints of ecologists, have persuaded us to examine most carefully
a wide variety of alternative control measures, old and new. These
include the use of parasites and predators (a principle well established
many years ago: see Chapter 20), the relatively novel idea of releasing
sterilized males, or males of a strain genetically incompatible with the
local variety, and recently introduced insect development inhibitors.
This is a large subject (see Chapter 20) and all that can be said here is

that none of these measures shows any sign of effectively replacing the immense beneficial effects of pesticides, although they may contribute to reducing their usage, in special cases.

Sensitive and perhaps emotional individuals in temperate climates may call for the abolition of pesticides; but the double needs of protecting crops and reducing disease transmission will override these anxieties for most people. Pesticides will continue to be used, but extra vigilance is needed. In particular, the possibility of the development of a resistant strain must be borne in mind when plans are made for any widespread usage. There are some reasons for believing that resistance is more prone to develop towards complex compounds with highly specific physiological action than to simple, non-specific, 'crude' poisons. This weakness of highly complex chemicals is not realized by those urging the use of specific poisons to avoid harming beneficial or non-target organisms. Above all, there is ample evidence that widespread and repeated heavy use of pesticides tends to select out resistant strains more rapidly than sporadic treatments. As a corollary, resistance can be prevented, or at least delayed, by discriminative use (in timing and placing) so that the maximum effect is achieved with minimum selection of the total population. Such careful use will also help to reduce environmental residues, if any.

If, despite all precautions, resistance develops to an intolerable level, the only practical answer at present is to change to a different type of pesticide. In the field of entomology, this worked well in the early days of DDT and other chlorinated insecticides. But, unfortunately, long-continued use of insecticides on a large scale has tended to accumulate varied defence systems in resistant strains. Cross-resistance is growing and the number of feasible alternatives is declining. This is exacerbated by the lack of commercial incentives to discover new pesticides, each of which costs several million pounds sterling to develop, largely on account of the extensive safety tests now required. There is no prospect of reducing these safety requirements, but perhaps patent legislation could be changed to allow a longer period of recuperation of these initial costs.

Pesticides should always be used in combination with any other feasible control measures. Some insect disease vectors can be largely banished from human dwellings by improved housing, water-borne sewage systems, and better general drainage and refuse disposal (all of which are at present too expensive for many of the developing tropical countries). Control of the vectors of these diseases can be supplemented by prophylatic drugs or immunization. For pests affecting crops and domestic animals, various agricultural practices are employed (crop rotation, changes of sowing, harvest or irrigation times) and the use of immune varieties.

Briefly, then, there is no single solution to the problem of resistance: each case must be considered from many angles. The pest control expert of the future will have to be widely knowledgeable to make the best use of integrated control systems appropriate to particular problems.

## Further reading

1 BROWN, A.W.A. and PAL, R. Insecticide resistance in arthropods. *Monograph Ser.* **38** (1971).
2 BUSVINE, J.R. Insecticide resistance and developments in pest control. *PANS,* **14,** 310-28 (1968).
3 ——, The biochemical and genetic basis of insectiside resistance. *PANS,* **17,** 135–46 (1971).
4 ——, and PAL, R. The impact of insecticide-resistance on the control of vectors and vector-borne diseases. *Bull. Wld. Hlth. Org.* **40,** 731–44 (1969).
5 FAO *Rep. 9th Session FAO Working Party on Pest Resistance to Pesticides.* FAO, Rome (1973).
6 WHO. Insecticide resistance and vector control. 17th Rpt Expert Committee on Insecticides. *Tech. Rep. Ser. Wld. Hlth. Org.* **443** (1970).

# 17 Pesticides and the conservation of land and energy resources

by D. Price Jones

The psychologist Maslow, in a contribution much cited in western educational circles, defines human needs in terms of values or 'metaneeds' such as truth, goodness, beauty, justice, order, self-sufficiency, and relates these to basic physiological needs. What emerges strongly from such considerations is that all that we regard as the higher activities of man are inescapably dependent on food, land, and energy, not just for their existence, but even for the manner and colour of their expression. Given only marginal access to these basic resources (a situation that exists today in certain parts of the world), man's cultural life approaches the edge of the gutter; indeed, it may be argued that, given the justifiable aspirations of man, such conditions are fundamentally worse than those experienced by a wild animal responding naturally and healthily to its own changing environment.

Political systems, inevitable products of man's social development, may express some of his cultural needs, but they, too, rest on the same basic resources: food, land, and energy. Land is the territorial imperative for a nation, whatever its political complexion. It is the home base and, with or without the aid of an exclusive language, it is a vital part of the national identity. As the arena for most of the nation's activities, it soon becomes a scarce resource. Agriculture, the producer of food and the major user of land, is required to become more efficient, to produce more food on less land, to release land for other uses.

In recent years, the management of scarce resources has become a major study area in official and academic circles, sometimes prompted by forebodings about the future. Of the numerous attempts to establish world system models, that of Meadows *et al.*[9] is perhaps the best known. In addition to incorporating land use, particularly in relation to food production, these models have become increasingly involved with energy resources. Although temporarily influenced by oil price rises, the real interest still attaches to the conservation of energy resources for the future benefit of mankind, particularly as energy itself is an

essential agent in the release of other scarce resources for human
consumption.

Pesticides have strong links with land utilization both through their
effect on the production of food and other commodities and, in their
public health application, the release of certain areas for profitable and
enjoyable human activities. The connections with energy lie partly in
the ecological energetics of crop protection and partly in the energy
cost of the benefits of pesticide use.

## Conservation of land resources

Pesticides can affect the management, and through it the conservation,
of land resources in several different ways:

(1)  by rendering the land fit for human habitation, through the
     control of pest-borne human diseases;
(2)  by controlling the pests of domestic cattle on, for example,
     land suitable for grazing but not for cultivation;
(3)  by increasing agricultural production per unit area, thus
     releasing land for other purposes, or compensating for land
     losses already incurred;
(4)  by protecting that production in the period between harvest
     and consumption;
(5)  by conserving the productive soil in the face of wind or water
     erosion.

These land-resource effects of pesticides are briefly reviewed below.

### *Control of human and animal diseases*

Certain human diseases have, or have had, an important — sometimes
an overriding — influence on land use. These include trypanosomiasis
(sleeping sickness), malaria in its various forms, yellow fever,
onchocerciasis (river blindness), the causative agents of which are
transmitted by insects, and schistosomiasis (blood fluke disease),
caused by a flat-worm associated with a water snail. (Chapter 3 deals
with vector-borne diseases in greater detail.)

It has been suggested that trypanosomes, transmitted by tsetse flies
*(Glossina* spp*),* delayed Bantu migration into Northern Transvaal before
1500 A.D., impeded Arab penetration of Africa from the east coast,
and also influenced the migration routes of Europeans travelling
northwards from the Cape.[2] Now, in the second half of the twentieth
century, trypanosomiasis still closely affects some millions of people
over nearly 4 000 000 square miles (10 000 000km$^2$) south of the

Sahara.[5] Over most of the area a considerable measure of control has
been achieved but much remains to be done. Occasional recrudescences
occur in areas where eradication had ostensibly been achieved, and
there are many areas where infection has remained, although at a low
level.

Control measures include spraying of riverine scrub with insecticides,
clearing the vegetation with herbicides to destroy tsetse fly habitats
around new settlements, as well as medication of the human population
and sometimes of domestic cattle. Resettlement of the land in newly
cleared areas is now considered essential: only normal human activity
can effectively and economically continue to suppress vegetation of
the type favoured by tsetse flies. Spraying of barrier zones to restrict
re-entry by the flies is also practised.

Mosquitoes transmit a number of diseases of which malaria is the
most important. As recently as 1957, there were about 250 million
human cases a year, deaths being estimated at more than two million.[10]
While this cannot be expressed in cold economic terms, observers
in the field have repeatedly drawn attention to one important effect:
the inability of infected populations fully to exploit the productivity of
their land. Eradication campaigns sponsored by WHO have achieved
near-miracles in reducing the incidence of the disease, even if recent
appraisals of the prospects of complete eradication have been a little
less optimistic. The use of pesticides, especially persistent pesticides,
has been an integral part of such campaigns and will continue to be so
for the foreseeable future.

Yellow fever was undoubtedly responsible for the retarded economic
development of parts of Africa and also the Caribbean region to which
it was introduced. The Panama Canal episode is the much-quoted
example. Nowadays, the chief vector, the so-called yellow fever
mosquito *(Aedes aegypti)* is susceptible to control by insecticides, but
additional measures, including modification of habitat, are necessary.
The disease itself is contained by human inoculation and restrictions
on travel.

Onchocerciasis, caused by nematodes (roundworms) transmitted by
blackfly *(Simulium* spp*)* is responsible for blindness in a high
proportion of the population in parts of West Africa, mainly in the
Upper Volta and neighbouring territories, but it also occurs elsewhere
in Africa and in Guatemala in Central America. The fly is associated
with the rivers (the larvae develop in or near water) and the human
population depends on the same rivers for its water supplies, including

water for its crops. The effective development of the land is therefore dependent on backfly control. Pesticides have proved effective in many areas but there are formidable application problems to be overcome in some other areas.

The blood-flukes (*Schistosoma* spp.) occur in many parts of Africa, the Middle East, the Far East, and in the north-eastern part of South America. They spend part of their life-cycle in water snails which feed on vegetation on the edges of ponds, lakes, rivers, and irrigation channels. The disease, schistosomiasis, is therefore common in irrigation areas. Blood-fluke eggs found in mummies testify to the high incidence of infection in Egypt 5000 years ago. The disease is extremely debilitating and offers serious obstacles to social and economic development. It has recently been estimated that some 590 million people per annum are at risk, with 125 million actually infected and 2.5 million entirely disabled; the total economic loss is put at about $640 million. New irrigation schemes in endemic areas have to make provision for snail control. Fortunately, suitable pesticides are now available.

These examples should suffice to show that the management of land resources is dependent on the health of the population and that pesticides have an important role to play.

*Increasing agricultural production*

The development of western civilization during the past two centuries has been accompanied by — indeed shaped by — increasing industrialization and urbanization. These processes have inevitably resulted in the transfer of land from agriculture to other uses. The potential loss of production has been made good partly by the reclamation or improvement of hitherto agriculturally unproductive land, but mainly by progressive improvement in agricultural technology. This improvement has accelerated during the past 30 to 40 years and still continues at a high level, although there is much discussion about future trends. Many of the technological inputs (fuel energy inputs, for example) have been concerned with increasing output per man, or reducing the physical work load; others — and fertilizers and pesticides rank high among these — have been directed more at increasing yields and conserving the resulting production.

Many attempts have been made to assess the contribution made by pesticides to increased food production but this is inherently difficult to measure. Individual experiments may give accurate data on yield increases but these cannot readily be translated into national or world averages. A courageous attempt to collate world *grain* losses due to pests, diseases, and weeds has been made by Cramer[2] and, although the situation is continuously changing, these estimates at least serve to indicate the possible scope for pesticides.

For present purposes, the estimate provided by the U.S. President's Scientific Advisory Committee[9] for the world-wide relationship between unit area production and pesticide input (both on a national basis) may serve as a guide (see Fig. 17.1). The curve presented is similar in shape to that for fertilizers and, one suspects, for agricultural technology as a whole, making a strict quantitative interpretation unwise. It is better perhaps simply to note that increasing pesticide

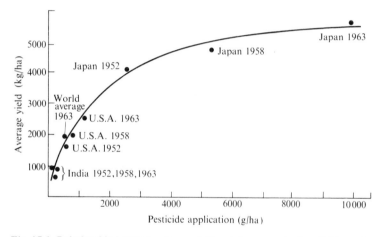

Fig. 17.1. Relationship between average yield and the amount of pesticide used per unit area for three countries with contrasting agricultural economies and for three years, 1952, 1958, and 1963.

input is associated with increasing production until at some high, but indefinite, level of pesticide input the curve approaches a plateau.

Fig. 17.1 may be interpreted in the light of data for crop areas in different parts of the world, and of the great variation in the level of pesticide usage. The inference is obvious: there are many regions of the world where pesticides used with discrimination on an increasing scale could result in greatly increased production. Some quantification of this relationship has been attempted by a research group at the Massachussetts Institute of Technology.[8]

As is fully discussed in Chapter 14, agricultural products are subject to deterioration between harvest and consumption. Hall[5] reports that losses may be as high as 50 per cent and in some cases higher, and implies that pests are largely responsible. Such losses are greatest in the warmer countries, which, all too often, are those with economies most in need of development.

## *Control of soil erosion*

According to FAO[3] the world's agricultural land in 1965 covered some 6000 million hectares and nearly all of it was in need of some protection from water. Extensive areas of badly eroded soil occur, for example, in India, the southern U.S.A., and central China. In contrast, the Japanese, with their intensive agriculture, have practised erosion control since time immemorial. The classical examples of wind erosion — deserts — are found in North Africa, the Near East, and in the arid and semi-arid parts of North America.

Husbandry techniques developed by agronomists and soil scientists have made soil erosion control feasible even in the highly mechanized agriculture of today. Nevertheless, the adoption of minimum cultivation systems, assisted by the use of certain herbicides, has simplified procedures and imparted a great impetus to the movement. It now seems probable that such techniques, which can usually be readily adapted to local conditions and requirements, could largely replace traditional cultivations, especially ploughing.

## The conservation of energy

The sun is the prime source of energy for life on earth. Solar energy trapped by green plants is used for the synthesis of a variety of materials which may, in due course, be eaten by herbivores, and these in turn by carnivores. Dead plant and animal materials are eventually broken down by detritus feeders and various organic and inorganic cycles are completed.

Ecological energetics is concerned with the flow of energy through different ecosystems and with the production of energy balance sheets. The paths along which energy flows are of interest because their relative importance varies with the species structure and with seasonal factors. In addition to the huge inflow of solar energy and the almost as great outflow of heat, energy may move in or out of any given ecosystem in the form of living or dead material borne by wind or water. Agricultural ecosystems have three outstanding energy characteristics: energy is diverted, as far as possible, into channels favourable to man; energy-rich material in large quantity is cropped and removed from the system; and large inputs of fossil energy are injected. Crop pests and diseases tend to divert the energy from the paths selected by man. The role of pesticides, and of other crop protection methods, can therefore be seen as the regulation of energy flow in agricultural systems in the interests of man.

*The energy input in pesticides*

The manufacture of pesticides is largely based on raw materials derived from fossil fuel, mainly oil. The initial energy debit is therefore the energy value of the oil, together with the energy cost of extraction and transportation (a convential addition that has to be applied with care). Other energy inputs go into the manufacture of the active ingredient, including the energy associated with other reaction materials and an appropriate item for the energy cost of the plant and its maintenance. In this way it is possible to arrive at a gross energy requirement (G.E.R.) for the pesticide active ingredient. To this must be added the G.E.R.s for the formulating materials and the container, as well as the energy cost of the formulation process. Then comes transportation by road, rail, or ship to depot and thence to the user: again, energy is consumed. Finally, the application of the pesticide (as a spray, for example) requires energy for driving the tractor and pump – and, of course, the equipment used required energy in its manufacture. The sum of all these constitutes the total G.E.R, which is generally calculated in megajoules (MJ) per hectare.

The energy requirements of pesticides are often difficult to calculate and they vary greatly for different active ingredients and different conditions of manufacture, formulation, transport, and use. Some very broad generalizations are therefore necessary and these are reflected in the adoption of 'notional' values. Similar generalizations and notional values apply to energy assessments for other methods of crop production and protection,[11] and the author of this chapter advances notional estimates for total G.E.R.s at the point of application;[7,8] these range from about 75 to about 450 MJ/ha.

*Energy balance in pesticide use*

Recent developments in integrated control notwithstanding, pesticides continue to be essential instruments for crop production, and their energy relations require examination. Insecticides and fungicides are used primarily to increase production, although quality is also a consideration in some crops. In crops for human or farm animal

consumption, the energy value of an increment of yield can be estimated
in nutritional terms, for example, as *metabolizable energy* (M.E.). On
this basis it is possible to establish the yield increments required to
balance a given pesticide total G.E.R. input. This has been done in
Table 17.1.

Table 17.1. *Increment of yield required to give an energy return equal
to that of a given pesticide energy input**

| Crop | Metabolizable energy (M.E.)† (MJ/kg) | Increment of yield providing 217 MJ M.E. (kg/ha) |
|---|---|---|
| Wheat | 15.0 | 15 |
| Barley | 13.1 | 17 |
| Potatoes | 3.2 | 68 |
| Sugar beet | 2.6 | 82 |
| Apples | 1.9 | 113 |
| Cabbage | 1.2 | 185 |
| Carrots | 0.96 | 226 |
| Celery | 0.33 | 654 |

\*  A notional input of 217 MJ/ha is adopted: see text.
†  Metabolizable energy is that part of the chemical energy contained in the food
   which can be used for growth and activity.
Source: *Outlook on Agriculture* (ICI)

The notional value for the pesticide input is believed to be rather
low. However, even if the value were doubled or trebled, it is clear
that, in energy terms, the required yield increments for grain, potatoes,
and sugar beet are well within the capabilities of modern insecticides
and fungicides. On the other hand, those crops of low calorific value
do not necessarily give a commensurate return in energy terms; their
contribution to human diet is not a matter of calories.

Similarly, if fibre crops such as cotton are examined, the fibre
itself (the main production objective) is not used for nutritional
purposes; if an energy comparison has to be made, it must be with the
energy cost of artificial fibres, and cotton seed production should be
used in the equation. In that case it is necessary to consider the energy
balance for the crop as a whole. Such an exercise serves to show that
balance sheets of this kind are not always meaningful unless set in a
very broad context.

Herbicides can be treated rather differently from insecticides and fungicides. For the most part, chemical weed control has been used as a replacement for more traditional hand or machine cultivations. Hence, a direct comparison between herbicide practice and older, but still current, practices, is both meaningful and at least theoretically feasible. However, cultivations vary in their weed control efficiency and, typically, several different kinds of cultivation may be integrated into a system. Furthermore, the systems — and their efficiency — vary from one farm to another and from one season to the next. What is more, a critical comparison should include all the energy costs, including that relating to the tractor driver.

These requirements pose enormous problems. A solution has been attempted by erecting standards for each type of cultivation and each energy input and combining these in a few tentative and illustrative systems. The exercise was undertaken for the sugar beet crop[8] but the principles are widely applicable. Only the conclusions can be presented here. Briefly, chemical weed control in general compares favourably with mechanical cultivations and represents a big saving in energy when ploughing can be replaced by the use of herbicides (see Table 17.2).

One energy input, that of the tractor driver, is difficult to evaluate but should not be neglected. Human labour can represent an insidious drain on the energy resources of the community but its evaluation and interpretation would take the discussion too far into the realm of socio-economics.

Table 17.2 *Energy inputs in different cultivation systems*

| Cultivation system | Estimated fuel consumption* | | Fuel energy input |
|---|---|---|---|
| | (Imp. gal./acre) | (l/ha) | (MJ/ha) |
| Traditional | 4.16 | 47 | 2150 |
| Reduced | 2.91 | 33 | 1500 |
| Direct drilling | 0.66 | 7.4 | 340† |

\* Based on information from U.K. Agricultural Development and Advisory Service.
† To this should be added the energy equivalent of the herbicide used, in this case Paraquat, for which, at 1.12 kg ion/ha, the additional energy would be about 855 MJ/ha.
Source: *Outlook on Agriculture* (ICI)

### Pesticides in agricultural and national energy budgets

Early studies of energy utilization in agriculture showed that improved yields were obtained at the expense of disproportionate increases in the energy employed. There were even some suggestions that advances in agricultural science and technology, other than mechanization, were unimportant. Such doubts have now been dispelled, but there remain in some quarters genuine misgivings about the apparently profligate use of energy in agriculture in the advanced countries, when the world resources are now seen so clearly to be finite.

However, in the U.K. the energy input into agriculture is only about 4 per cent of the total national consumption. Much more (about 6.5 per cent) is used in the food-processing industry and in distribution. A roughly similar relationship holds in most Western European and North American countries, although the position is affected to some extent by variations in imports and exports. It follows therefore that whatever energy savings are made in agriculture, they cannot possibly make a big difference to national energy budgets in those countries.

Similar considerations apply to pesticides as an input into agriculture. In the U.K. the pesticide-related energy input is probably less than 2 per cent of the total energy input in agriculture; in the U.S.A. it is about 2 per cent for the corn (maize) crop and of the same order for agriculture as a whole. Again, it is obvious that restrictions on the use of pesticides cannot possibly make significant savings in the use of national energy resources.

*     *     *

Some of man's resources, land and energy among them, are finite and his use of them demands constant reappraisal. His food comes mainly from the land, which is already in short supply in many parts of the world. Food production is making heavy demands on yet another scarce resource — energy. Pesticides, used intelligently, can help to conserve some of these scarce resources. Land can be rendered habitable, its productivity increased, and the production safeguarded; the soil itself can be defended against wind and water. Pesticides likewise ensure that the energy cost per unit of production is reduced, mainly by contributing to higher production on a given area; but partly — and perhaps increasingly, as in certain weed control operations — by reducing the energy input into the crop.

### References

1   BENNETT, I.L. (Chairman) *The World food problem. Report of the Panel on the World's Food Supply. President's Scientific Advisory Committee.* Govt Printing Office, Washington, D.C. (1967).

2   PUXTON, P.A. *The natural history of tsetse flies.* H.K. Lewis, London (1955).
3   CRAMER, H.H. Plant protection and world crop production. *PflSchutz-Nachr. Bayer,* **20,** (1) (1967).
4   FAO. Soil erosion by water: some measures for its control on cultivated lands. *Agric. Dev. Paper FAO.* **90,** FAO, Rome (1970).
5   FAO/WHO African trypanosomiasis. *FAO Agric. Stud.* **81;** *Rep. Joint FAO/WHO Expt Cttee,* FAO, Rome (1969).
6   HALL, D.W. Handling and storage of food grains in tropical and sub-tropical areas. *Agric. Dev. Pap. FAO* **90,** FAO, Rome (1970).
7   JONES, D. PRICE. Energy considerations in crop protection. *Outl. Agric.* 8 (3) 141 (1975).
8   —— , Herbicides as an energy input in sugar beet production. *Proc. 3rd Int. Meeting on Selective Weed Control on Beet Crops, Paris, 1975.* I.I.R.B., Tirlemont, Belgium (1975).
9   MEADOWS, D.H., MEADOWS, D.L., RANDERS, J. and BEHRENS, W.W. *The limits to growth: a report for the Club of Rome's project on the predicament of mankind,* Earth Island, London (1972).
10  PAMPANA, E.J. and RUSSELL, P.F. Malaria: a world problem. *Chronicle Wld Hlth Org.* **9** (31) (1955).
11  *Span* special issue: Energy and agriculture. *Span,* **18** (1) (1975).
12  WILSON, C.L. (Director) *Man's impact on the global environment. Report on Study of Critical Environmental Problems.* M.I.T. Press, Cambridge, Mass. (1970).
13  WRIGHT, W.H. A consideration of the economic impact of schistosomiasis. *Bull. Wld Hlth Org.* **47,** 559 (1972).

# 18 Pesticides, the environment, and the balance of nature

by K. Mellanby

Much of this book shows how pesticides protect crops from attack by pests. Crop protection has generally been successful, and, in many cases, has proved to be an essential part of modern agriculture. Nevertheless there is widespread concern that the environment in which we live may be being damaged — polluted — by these very chemicals, and that something generally spoken of as the 'balance of nature' may be being upset in a manner harmful to man and to the whole globe on which he lives. It is my purpose in this chapter to consider the validity of these misgivings.

The whole concept of the 'balance of nature' needs to be examined. Ever since life existed on earth, it, and the environment generally, has been changing. Evolution has proceeded. New forms of life have appeared, and have altered the world in various ways. The vast majority of species of plants and animals are already extinct, our present flora and fauna consisting of those which, in the conditions obtaining during the last million or so years, have been able to survive. Evolution continues, and in time it is probable that different animals and different plant species will replace those with which we are today familiar. How long the human species itself will survive is an open question. Change rather than balance seems to have been the order of the day.

Nevertheless, before man dominated the earth there was some degree of stability, which lasted over short periods of perhaps several thousand years. Roughly the same sort of number of the different species of plants and animals coexisted year after year in each area — in what might be spoken of as a 'natural ecosystem'. In such ecosystems there was great diversity, that is to say there were many different plants and animals. Some were rare, some were common, and though absolute numbers changed, as a rule a recognizable pattern was maintained. Observations of this kind have given rise to the belief (a belief often elevated to the status of a dogma) that 'diversity produces stability'. There is some truth in this belief, but the truth is not an absolute one. Even in completely natural areas, unaffected by man, population

Plate 18.1. Plants that to the naturalist are beautiful wild flowers may be weeds in the eye of the farmer. This pasque flower (*Pulsatilla vulgaris*) is an example of a plant that may become rare in some temperate countries through the use of herbicides to improve grassland. (Photograph by M.C.F. Proctor)

explosions of locusts or lemmings can drastically upset the balance for a time. But generally something like stability is eventually restored.

Incidentally, it may be asked why should we worry when some species of wild animal is exterminated by man, whether by using pesticides, by destroying the habitat, or by over-shooting, if so many more extinctions have taken place without his intervention. The answer lies in the speed with which these processes occur. When a wild species, say a primitive horse like *Eohippus,* died out, this was in part because it was replaced by a new species better adapted to the changed conditions. When man killed off the last Dodo or Great Auk, there was not time for any replacement to be evolved and the world was permanently impoverished. Man-made changes are too rapid to allow evolutionary processes to operate.

### The pattern of modern farming

Man has drastically altered the whole situation in the last ten thousand years, mainly by his adoption of arable agriculture. Before that, he was just one animal among many others. His numbers were small, for the world cannot support many hunter-gatherers. It has been suggested that the stable human population for Britain under these circumstances was only about 5000. Arable farming changed everything. It so increased the productivity of the land that it could eventually support a population increase in the order of 10 000-fold, and this made civilization and urban life possible. It also gave rise to many new problems.

An arable crop – an area of monoculture of one type of plant – is the very antithesis of a natural ecosystem. It is obviously completely lacking in diversity. Ecological principles would make one expect great instability in such an area, and this in fact often occurs. This instability is demonstrated by vast and rapid increases in the population of insects which, outside the crops, can only be rarely found. Such insects in these numbers may do great damage to the crops which have allowed them to spring up. They are, in fact, pests.

However, arable farming does not just produce vast areas of monocultures of crop plants. It, and its attendant urban growth, alters the entire environment. This upsets a whole series of natural balances. The results are not always harmful, even to wild animals and plants which are often thought particularly to be at risk. When Britain was covered with forest in the days when the few human hunters lived, the countryside must by any reckoning have been very dull and monotonous. Some species of large mammal, such as the bear, the wolf, the elk, and the beaver, which are now extinct in Britain, were present, but many of our common and more attractive birds, such as the robin, the blue tit, and the song thrush, must have been comparatively rare. Our much-admired meadow flowers were restricted to patches of ground where trees had fallen and woodland regeneration had not yet produced a closed canopy of branches. In many ways, wild life gained from the changes caused by farming, which often in fact created, rather than destroyed, large-scale diversity.

Nevertheless, farming produces conditions which favour the development of pests, a term I use to include plants generally thought of as weeds. A large part of the work done by the farmer in the past was aimed at their control. Much soil cultivation was to reduce the number of weeds; crop rotations played a part in reducing pests and plant diseases. Pesticides have been developed to do things farmers have been trying to do for generations by other means, and to do them more efficiently. As other chapters show, they have very generally succeeded. Thus, when herbicides make it possible for a field of wheat to be grown with no charlock or wild oats, the result is similar to that produced by an earlier generation of farmers but with the expenditure of a great many more man-hours in the process. Why then should anyone be concerned?

### Why pesticides are criticized

Pesticides are criticized, particularly by environmentalists and those concerned with wild-life conservation, for two reasons. First, although (as has just been shown) they do things which farmers have been trying to do by other means for generations, they do so much more thoroughly and efficiently. This, as will be shown, has some unexpected effects. It sometimes kills off harmless or even beneficial species, which may previously have been overlooked. It also makes possible large-scale changes in the landscape which were previously difficult or even impossible to achieve. Sometimes this has allowed ill-advised policies to be implemented before their total effect has been understood. The second criticism is that pesticides have had unwanted, and often unexpected, harmful side-effects, sometimes arising from misuse, but also on occasion because of the chemical and physical properties of the chemicals themselves.

### Are pesticides too efficient?

As already mentioned, cereal crops can now be grown more cleanly than ever before, with next to no weeds to contaminate them. Many weeds are harmful, reducing yields by competition for light and nutrients, or by making harvesting more difficult. However, there are also some weeds commonly found in cereals, plants like violas, which do not appear to harm the crop at all. Now in many countries shooting of game birds is of some economic importance, adding (by the sale of shooting rights) to the income of the farmer or landowner, or even providing him with some well-earned recreation. In recent years numbers of the grey partridge have been very low, and the bird has virtually disappeared from parts of Europe.

In England, this fall in partridge numbers has been shown to be caused by the use of herbicides. There has been no direct harmful effect of the chemical on the birds, but the elimination of small weeds has reduced the population of small, quite harmless insects which live on these weeds. Young partridge chicks in May and June normally obtain a substantial part of their food from these insects, and, particularly in unfavourable wet and cold weather, the cutting-off of this food supply is sufficient to increase chick mortality to catastrophic levels. The reduction of permanent grass, and of the practice of undersowing cereals, also adds to the shortage of insects to feed the partridge chicks, but the unnecessarily clean crops seem to be the major factor.

In other countries, including West Germany, where weather conditions may be different, the fluctuation in partridge numbers does not appear to be correlated with the use of herbicides. So far we only know for certain that partridges in England are being affected, because this species is of economic importance and accurate records of game shooting are kept. There are, however, suggestions that wild species of birds may be suffering in the same way and that this may explain some fluctuations in numbers. Farmers interested in partridge shooting are certainly trying to be just a little less 'efficient' in their efforts to control the less harmful weed species.

There is a general impression in Britain, and in parts of the United States, that many types of butterflies are much less commonly seen today than they were thirty years ago. There is good evidence to show that some species are indeed rarer, and that sites in which they existed have become deserted, though other species have increased in numbers and extended their distribution. There is a widespread belief that where butterflies have decreased 'agricultural chemicals' are to blame.

There is little evidence to show that insecticides have had much effect on butterfly numbers or populations, except where pest species, such as the cabbage white, have been caught as caterpillars *in flagrante delicto* eating cabbages, and have been controlled within the crop by the proper use of the appropriate chemical. Serious reductions of desirable species have usually been caused by the loss of the caterpillar's foodplant, or by the loss of some particular feature in the habitat.

Nettles are usually considered to be noxious weeds, and their easy control by 2,4,5-T and other herbicides is generally welcomed. However, several of the commoner and more beautiful butterflies, including tortoiseshells, peacocks, and red admirals, which give so much pleasure to gardeners, have nettles as their larval foodplant. The cutting of road-side verges, the tidying-up of rough areas around farms, and the efforts of gardeners themselves has undoubtedly reduced the area of nettles. Now that the importance of this plant for butterflies, and as a refuge during the winter for several beneficial insects which prey on pests, is being realized, conservation-minded gardeners are glad of the excuse to leave some patches of these weeds, getting a feeling of virtue and doing less work into the bargain.

The habitat most endangered in Europe today is old permanent grassland. Some areas on the chalk have remained uncultivated for at least two thousand years, and many water meadows are of considerable antiquity. This grassland is generally very rich in species, including 'weeds' of considerable botanical interest. Unfortunately such species-rich herbage is relatively unproductive, either as grazing or for hay or silage. Productivity can be improved by using selective herbicides to eliminate all but the grass, and by dressing with fertilizers which stimulate the grass and suppress other plants. The results are just what the farmer requires, but he obtains them with the loss of diversity of both plants and insects. Many species of plant and butterfly have disappeared from large areas of Britain as a result of this grassland improvement. There is no easy answer to the problem. It has been suggested that farmers should receive cash compensation for retaining important areas of grass in an unproductive state, and that more nature reserves, bought and maintained by conservationists either with public funds or those raised by voluntary bodies, should be declared. But this must be recognized as an area of real conflict between conservation and agriculture, and one made more real by the quite proper use of chemicals which have an indirect, rather than a direct, effect on species considered desirable by a section of the population. It is encouraging to find that some farmers have voluntarily adapted their methods on such areas to preserve their ecological interest.

### Rapid changes in large areas of land

In the past there have been two main farming revolutions in Europe and these have been best documented in Britain. The enclosures of the eighteenth and nineteenth centuries, with the introduction of new crops and new rotations and a great increase in the numbers of overwintering livestock, as well as the planting of the majority of the hedges, were part of a comparatively gradual process, although it was in truth a revolution. There was serious unrest, at least for a few years, in some areas, especially when workers were impoverished by the loss of common grazing rights, but the increase in agricultural productivity had many beneficial effects and the slow pace of change made it more easily assimilated. As a by-product of this first farming revolution, the countryside became more beautiful, more productive, and even richer in wildlife.

The second farming revolution, based on petroleum power and agricultural chemicals, has taken a much shorter time, having been almost completed in less than 30 years. It is possible to alter the whole appearance of vast acres in weeks or months, and to produce results which are not universally considered to be an improvement. For instance, there is no doubt that bracken is a weed, reducing the land's productivity and even poisoning stock. Farmers have often sought to eliminate it from their land, but until recently have only been able to treat comparatively small areas. The introduction of the herbicide asulam has completely altered the situation. Large areas can be sprayed from the air, and the bracken destroyed. Although asulam has been carefully tested and found to be of little direct danger to wildlife, this use has been widely opposed by conservationists who are anxious lest too great an area be improved, with the possible loss of both wildlife and visual amenity. This is obviously another difficult area of conflict.

The very efficiency of modern pesticides has also made it possible to give effect to mistaken and irreversible policies which are later regretted. For example, in Britain the Forestry Commission has suffered in the past from changes wished on it by different governments. At one time it was expected to concentrate all its efforts on timber production, without regard to amenity, wild life, or other users of the countryside. Many of its officers resented this lack of consideration for others, and did a good deal to try to mitigate its results. There has eventually been a considerable change, and though efficient timber production is still required, the use of woodland for recreation and wild life is now acknowledged. Lowland areas are now being planted more often with native hardwoods instead of regimented rows of exotic conifers.

The present situation in lowland Britain would be better than it is had not chemical herbicides been so efficient. Many areas of conifers had been interplanted with hardwoods, especially oak, with the idea that when the quicker-growing softwood had been harvested the oaks would survive and continue to grow into a more 'natural' forest. Unfortunately, in the 1960s a policy decision was taken to remove the oaks, probably to allow a second rotation of conifers when the first were cut. Without herbicides, it would have taken years of hard work to cut out the oaks and many would probably have survived. But it was discovered that 2,4,5-T, applied by air at the right time, could kill the oaks without harming the evergreens. Several thousand hectares were treated before the policy was reversed. By that time the oaks were dead.

### Pesticides and beneficial species

A number of important pests, particularly of perennial crops like citrus trees, have been controlled by the encouragement of their natural enemies, a process often spoken of as biological control (see Chapter 20). If a second pest, not so controlled, appears, then its control by chemical means is likely to produce a recrudescence of the first. This happened in California, where the cottony cushion scale had been successfully controlled since late in the nineteenth century by a ladybird beetle introduced from Australia. Recently the scale has once more caused damage because its predator has been inadvertently damaged by insecticides used against several other pests of the citrus trees.

Biological control is, of course, the scientific use of a process occurring all the time in all habitats. The ecological stability mentioned earlier in this chapter depends, in part, on the control of the growth in numbers of each species in an ecosystem. A predator or a parasite prevents its victim from increasing its numbers to anything like its potential. In the simple ecosystem of a monoculture, pests may develop because predators and parasites are absent. Pest damage is generally less serious in Western Europe than in countries with a shorter agricultural history, because some degree of equilibrium between the pests and their enemies has been established. Pesticides may sometimes upset this equilibrium.

The most striking case of this occurs in apple orchards. Red spider mites have been known for many years to be present but not in sufficient numbers to do economic damage. Moths, whose caterpillars damage buds and fruits, aphids, and other insects were destroyed as eggs by winter sprays of tar distillates after 1921. These sprays also destroyed the predators of red spider mites, which then could cause serious damage. DDT had the same effect in forests in the U.S.A. The remedy is to use a different insecticide or to have an extra spraying with an acaricide. Thus, some insecticides produced what was, in effect, a new pest requiring control.

Entomologists are well aware of the dangers referred to in the last two paragraphs, and chemists try to produce selective insecticides which are less dangerous to beneficial insects. Most progress has been made in carefully timing the application of a spray to avoid the beneficial and to catch the harmful insect; this is the basis of 'integrated control' which attempts to make the best of both worlds. Unfortunately, a great deal of ecological knowledge about all species concerned is necessary before it can be really successful.

Resistance to various insecticides by many insects has been reported from countries in all continents. This is dealt with in Chapter 16.

## Unintentional damage by pesticides

There have been many occasions when pesticides have been carelessly or wrongly used, and when serious damage has been done to the environment and to human beings as well as to wildlife. When the auxin type herbicides were first introduced their potency was not always realized, and spray drift frequently damaged gardens and orchards near to a target. Spraying of insecticides and herbicides from aircraft is a highly skilled operation, and sufficient care is not always exercised by the operator. Very poisonous insecticides such as parathion are thought to have caused many deaths of insufficiently careful operators in the Far East and even in the U.S.A. Rivers, streams, and lakes have been polluted by the careless or irresponsible discharge of pesticide wastes, sometimes with disastrous and longstanding effects.

It is obvious that the more potent the pesticide, the greater the danger it may be to the environment, and the greater the need for care in its use. The generation of pesticides now being introduced appears, in general, to be less toxic, and less persistent, than some of those which came on the market nearly 30 years ago. These chemicals still have their uses, but it is likely that they will eventually be replaced with substances which are safer (though they may be more expensive and less effective).

## Pesticides as pollutants

Pollution occurs when a substance is introduced into the environment, generally by man, and damage is caused to life or to amenity. There is a widespread belief, particularly among non-scientists concerned with environmental problems, that pesticides, particularly DDT and other organochlorine insecticides, are causing widespread pollution. It is even suggested that the future of our globe itself is in jeopardy, that DDT may interfere with the oxygen supply and, eventually, make life impossible. There is no doubt that there is *some* pollution, and that pesticides have killed birds and other animals unintentionally. There is no doubt that man-made pesticidal chemicals can be detected, by chemical analysis, in almost every part of the world. However, there is disagreement about the seriousness of the problem, and my intention is to give, as briefly as possible, a balanced account of the situation.

Many workers have found traces of DDT and other organochlorines in rain, in birds and other animals, in snow, and in fish at sites remote from those where pesticides have been deliberately applied. It is probable that this global contamination occurs because the chemicals are volatilized into the atmosphere and deposited with rain. At present the levels arising in this way cannot be called pollution, for they are so low as to cause no detectable effects on living organisms. It has, however, been suggested in such publications as *The limits to growth* that, even if no more DDT were used anywhere in the world, the substance is so persistent, and is transported in such a way, that levels in the ocean are likely to increase for at least the next 30 years to a level which has been shown to affect living organisms and possibly to reduce the efficiency of the phytoplankton to recharge the atmosphere with oxygen.

Fortunately, there is evidence which allows us to discount this opinion. Sea-birds, feeding on fish in estuaries, are apparently good indicators of the presence of organochlorine compounds. The level in their body fat is related to, but much higher than, that in the estuary water. In 1968 levels were found to be increasing in shags and other species caught off both the European and North American coastlines. Since then use of the pesticide has decreased; so has the level in the birds. There is no evidence to support the view that doomsday is being brought nearer by pesticides.

Instances of local pollution, and the death of wild life, have been well documented. Persistent and stable chemicals, unlike many poisons, can act as secondary poisons killing animals which consume those which

226

have themselves ingested the insecticide. In some cases higher levels are found in members higher in food chains. Sub-lethal effects, caused by levels near to but lower than those causing death, have been described. But where such incidents have been carefully studied, as for example in the bird kills in Britain caused by aldrin and dieldrin seed dressing in the late 1950s and early 1960s, once stricter controls of insecticides have been introduced the process has been reversed.

Serious pesticide pollution has, however, occurred in one habitat: fresh water. Here levels in the region of one part in $10^8$, arising from effluents of factories engaged in moth-proofing or, less frequently, by run-off from farmland, have caused fish kills and death to many invertebrate animals. This is because the animals breathe the oxygen dissolved in the water and at the same time absorb into their bodies the pesticide contained in the enormous volume of water passed over their gills. The main reason for restricting the use of DDT and other such chemicals is that they may inadvertently get into fresh water, where they will almost certainly cause serious damage. If carefully used in other media they are much less dangerous.

Thus, it is clear that pesticides *can* be pollutants, that they *may* cause environmental damage, and that they *do* cause widespread but, apparently, harmless contamination. It is impossible to prove a negative, and to be certain that this does not add up to more danger than would, at first, appear. For that reason many people think it is wise to control the use of such substances. If millions of citizens in a developing country would die from malaria unless there was a fairly generous use of DDT, it would probably be right to risk some temporary effects on wild life. In a wealthy western country it may be sensible to use a more expensive and less efficient pesticide which is less likely to have side-effects. The danger is that if the western world bans persistent insecticides before adequate substitutes are available, less developed countries may follow their example with disastrous results.

**Further reading**

1. BEATTY, R.G. *The DDT myth: The triumph of the amateurs.* John Day, San Francisco (1973).
2. CARSON, R. *Silent Spring.* Houghton Mifflin, New York (1962).
3. GUNN, D.L. Use and abuse of DDT and dieldrin. In *Foreign compound metabolism in mammals,* vol.3, pp 1-82. Chemical Society, London (1975).
4. MELLANBY, K. *Pesticides and pollution.* Collins, London (1967).
5. MORIARTY, F. *Pollutants and animals.* Allen and Unwin, London (1975).

# 19 Pesticides: the legal environment

by R.F. Glasser

When one talks about the legislation and regulations which deal with pesticides in relation to crop protection and human welfare, it is of interest to know how they came about and how they provide protection to man and his surroundings in the context of the total environment.

Obviously producers of pesticide products must understand these laws and particularly those aspects which involve regulations and control of use of the products which they manufacture. The purpose of this chapter is to acquaint people outside the industry with the laws and regulations which are designed to protect them.

From the world-wide standpoint, legislative and regulatory aspects of pesticides are complex: each country has its own unique pattern of legislation, registration schemes, and their enforcement. This complexity is increased as a consequence of the large number of ways in which pesticides for plant protection and non-agricultural uses may possibly affect human well-being. Consequently, a number of different government departments in a country (public health, agriculture, labour, transport, etc.) usually have a particular interest in the legislative control of pesticides.

Pesticide laws have evolved from early poison legislation, where the pesticides had often been included in the lists of poisons annexed to such legislation. With the development of the agricultural chemical business and the increasing use of pesticides, national governments have strengthened laws, or introduced new legislation, to deal specifically with the control of their use. The main objectives of legislation are:

(1)   to protect people who may be exposed to acute risks during manufacture, formulation, packaging, transport, and storage;

(2)   to ensure good packaging, which should carry the proper classification as to risks; and to prevent direct contamination of food and animal feed at any point between manufacture or formulation and use in the field;

(3)   to protect people who may be exposed to risks when opening containers, diluting a concentrated chemical to obtain the correct field dosage, and subsequently applying the pesticide;

(4)    when appropriate, to warn against unintentional contamination of non-treated crops, animals, soil, and water;

(5)    to protect the buyer against the sale or purchase of low-quality products or against misleading claims made on the label or in advertising;

(6)    to protect consumers of the treated food or animal feed by ensuring that the pesticide is correctly applied and that adequate intervals between application and harvest are clearly established and, where appropriate, stated on the label. This ensures that pesticide residues, if any, remaining on the food or feedstuffs are of acceptable levels so that there is no risk to the consumer; and

(7)    to ensure that the consequences of the foregoing requirements do not place unwarranted restrictions on the development of new pesticides.

Thus it can be seen that legislation has been drawn up in such a way as to protect man from harming himself and, often more important, other people and their surroundings.

Most of the legislation enacted over the five years 1970-5 has entailed consultation with industry, and with a few notable exceptions such new legislation has been found to be workable. The laws, as one might suppose, vary widely throughout the world, but the regulations all reveal a striking similarity in the basic data they require. Perhaps the underlying difference between countries in respect of regulatory requirements is that some accept the scientific case and have the competence to assess the detailed data presented, while others adopt to a greater or lesser degree a strict legalistic approach.

Inevitably, the amount of information required for registering a new pesticide is increasing. How much broader the scope and complexity of this type of data will become is unpredictable since science is never in a static state. What is certain is that the cost of obtaining the scientific data to meet registration requirements is escalating alarmingly, and this is reflected in the reduction in the effort spent on research to find new pesticides.

## Harmonization

To those who work with regulatory authorities, harmonization of registration requirements (i.e., biological efficacy studies, labelling, packaging, toxicological investigations, residue trials, and environmental safety) is most desirable, and various organizations are involved in

229

harmonizing schemes. This view is being emphasized by the Food and Agricultural Organization of the United Nations (FAO) and has been supported by the pesticide industry through its trade association, Groupement International des Associations Nationales de Fabricants de Pesticides (GIFAP), for example during active participation in the FAO Ad Hoc Governmental Consultation on Pesticides in Agriculture and Public Health that resulted in a report published in 1975.[7] In this report it was agreed, under the heading 'Ad hoc government consultation', that legal registration of pesticides was needed, but that there were difficulties arising from the increasingly divergent national registration requirements, which were having the effect of increasing costs and tending to inhibit the development of these critically needed pest-control materials. It was recommended, and plans are under way, that FAO, in collaboration with the World Health Organization (WHO), should progress the harmonization of requirements for registering pesticides, and that those involved in national and international registration schemes, as well as representatives of the pesticides industry, should participate.

The Council of Europe is also involved in harmonizing schemes for the approval, classification, and labelling of pesticides. This body was established by ten nations in 1949 with the aim of harmonizing activities in economic, social, cultural, educational, scientific, legal, and administrative matters and in the maintenance and further realization of human rights. The two main bodies of the Council are the Consultative Assembly and the Committee of Ministers. The Assembly consists of Members of Parliament appointed by national governments. The Committee of Ministers is served by national experts for all 'technical' questions, and for pesticides by the Subcommittee on Poisonous Substances in Agriculture and the Ad Hoc Study Group on Pesticides (Fig. 19.1). Both groups, through their technical members drawn from the European national governments, focus their attention on the harmonization of pesticide regulation guidelines and programmes affecting environmental conservation. A monograph published by the Council of Europe [3] contains the representative views of these governments and of industry, and has been widely accepted as an example for drawing up or modifying guidelines for pesticide regulations in many countries.

So far as proposing maximum residue limits in food or animal feed is concerned, a subcommittee of the Council's Subcommittee on Poisonous Substances in Agriculture maintains liaison with the FAO

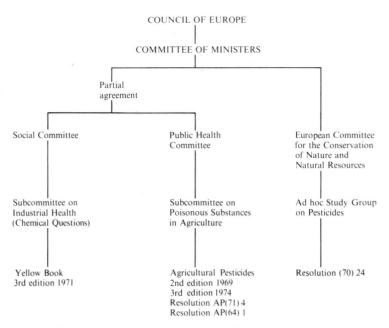

Fig. 19.1. The Council of Europe Committee and their publications related to pesticides.

Expert Committee on pesticide residues and generally accepts its recommendations.

The other major body in Europe looking into the harmonization of pesticide legislation is the European Economic Community. The concern here, however, is mainly the standardization of food regulations within its member countries. Commodities that are exported to countries outside the Community do not come under its terms of reference. In general, the proposed maximum residue limits are relatively low and reflect the use of products only within the community itself; they would not necessarily be acceptable in establishing permissible residue levels of pesticides in international trade.

Also of great importance is the work of the Codex Committee on Pesticide Residues, which operates under the *Codex Alimentarius* Commission of the United Nations (Fig. 19.2). In contrast to the two European bodies mentioned above, the Codex Committee on Pesticide Residues is truly international and, with FAO and WHO as expert consultants, is active in promoting and trying to harmonize residue limits for those pesticides which may be used on foodstuffs moving in world trade and which should not be exceeded (known as 'maximum residue limits'). The Committee is made up of representatives from governments of major food exporting and importing countries and meets at regular intervals. Important to the work of the Codex Committee on Pesticide Residues is the WHO-FAO Joint Expert Committee on pesticide residues, which is composed of scientists chosen for their wide knowledge and expertise in toxicology and chemical residues. As they are also familiar with the practical use of pesticides, they are able to evaluate the available residue data to determine maximum residue limits which may be recommended as being consistent both with good

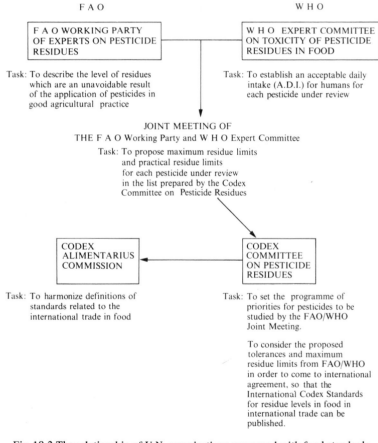

Fig. 19.2 The relationship of U.N. organizations concerned with food standards.

agricultural practice and with the acceptable daily intake proposed by the WHO experts.

It is clearly of great importance to food-producing countries that there should be world-wide acceptance of the maximum residue limits proposed by the Codex Committee. It is even more important that these limits are set at levels that will embrace the residue actually arising as a result of using the product according to official recommendations in producing countries.

**Typical legislative and regulatory procedures**

In terms of registration (approval for use and sale of a pesticide), the most demanding requirements on both industry and government are that both should be able to satisfy themselves that the product proposed for registration not only makes a useful and acceptable contribution to pest control, and thereby human welfare, but does so without the risk of unacceptable side effects from the viewpoint of safety, food residues, and the environment. In most countries an 'approval authority' normally obtains the bulk of its information from the pesticides manufacturers concerned. This may well be complemented by published data obtained by independent sources, such as FAO or WHO. All the data, however, are utilized in the approval procedure which ultimately results in a registered label.

The label on the pesticide container is a specially significant factor in pesticide legislation. It is the principal means by which the manufacturer and the authority instruct the ultimate user of the pesticide on how to obtain the best results consistent with its safe use. Agreement of the label text thus becomes a very important feature of legislative procedure in dealing with pesticide products.

Most countries now require in their legislation that information on the label should include the following :

(1)   a statement describing the nature and quantity of the active ingredient;

(2)   recommendations for use (where and when to apply, and the quantity to be used);

(3)   the minimum period that must be allowed to elapse before the treated crop may be harvested; and

(4)   the necessary warning or cautionary statements, which must be adequate both to prevent injury to the person handling the product and to protect the public and animals, including wild life such as birds and fish;

(5)     according to the assessment of the risks which may be associated
        with the use of the product, legislation may require 'symptoms of
        poisoning', 'directions for first aid', and 'information for the
        physician' to be included on the label. Such information is nearly
        always included on the container label for technical materials (see
        Appendix).

Many countries are including the establishment of maximum residue
limits in their regulations as a means of safeguarding the consumer. Most
countries also insist that pesticides, when used in accordance with the
recommendations on the label, should not leave unacceptable residues in th
foodstuffs. To appraise acceptable residue limits authorities take into accou
the relevant dietary habits, the toxicology of the chemical and its metaboli
and the decline of residues during post-harvest processing.

Maximum residue limits are then established which do not cause a risk to
public health. This approach takes into account the many factors of 'good
agricultural practice' that influence the occurrence of residues, such as the
pesticide itself, its formulation, the rate of application, method of applicatii
time of treatment, the number of treatments, and the interval between last
application and harvest – to name the most important considerations. By tl
end of 1975, some 25 major crop-producing countries had maximum residu
limits laid down in their regulations.

Toxicology matters concerning pesticides are usually handled by experts
in ministries of health. Toxicology data are evaluated to determine the
dosage rates at which toxic effects occur in test animals; for maximum
value they must then be extrapolated in terms of man. How these data
are used in regulating pesticides varies from country to country. Some
countries simply use them to guard against hazards to those who use
the product. Others, in addition to this, use toxicology data to show
that residue levels arising from recommended uses will not be harmful.

### Some typical regulatory schemes

The regulatory schemes for the Federal Republic of Germany, the
United Kingdom, Japan, and the United States of America illustrate the
varied but typical regulatory procedure activities. The Federal Republic
of Germany and the United Kingdom provide good illustrations of two
quite different approaches to regulating pesticides.

*Federal Republic of Germany.* In West Germany, legislation demands
that agricultural pesticides be registered with the Federal Institute of
Biology for Agriculture and Forestry (BBA). This entails the submission

by the manufacturer of a special application form giving detailed data on toxicology, residues, biological performance, etc. It is essential that biological and residue data should include results from official trials carried out in West Germany in accordance with guidelines laid down by the Government. The BBA checks the biological reports, the composition of the product, and its physical and chemical properties, while the toxicology and residue data are reviewed by the Federal Health Office (BGA), which considers the toxicity classification and, where applicable, official maximum residue limits.

In general, German maximum residue limits tend to be lower than levels set by other countries, which, of course, can pose problems in trade with countries exporting agricultural products to West Germany.

*United Kingdom.* In contrast, the existing regulatory schemes for agricultural pesticides in the United Kingdom are not governed by direct pesticide legislation; rather, they have evolved over many years from voluntary agreements between the pesticide industry and government health and agriculture authorities. There is a Pesticide Safety Precautions Scheme, which is concerned mainly with the clearance of safety aspects, including toxicology, residues, and environmental issues; and an Agricultural Chemicals Approval Scheme, concerned with biological efficacy. In the latter case, field trials are carried out by the manufacturer and these are open to inspection by officials of the scheme. Fig. 19.3 outlines the programmes for notification, clearance, and approval of a pesticide in the Federal Republic of Germany and in the United Kingdom.

Unlike Western Germany, the U.K. has no statutory maximum residue limits for pesticide residues, except for arsenic and lead. The official recommendation sheets issued for each product cleared under the schemes include, where appropriate, statements on the typical level of residues to be expected from use of the product according to the manufacturer's label recommendations. In this way, control of residue levels is accomplished by following the principal of 'good agricultural practice' instead of by enforcing statutory maximum residue limits. The Government Chemist does, moreover, carry out regular analyses of foods for pesticide residues known as 'total diet studies', and measurements have shown that residue levels, when detected, are usually well below the acceptable daily intake figures laid down by the FAO-WHO Joint Committee.

Although West Germany and the U.K. serve to demonstrate two

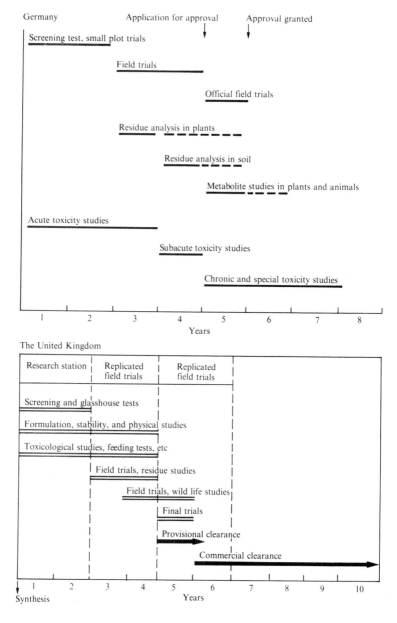

Fig. 19.3. Comparison of procedures for registration of a pesticide in West Germany and in the U.K. (notification, clearance, and approval).

entirely different methods of regulating pesticides in Western Europe, moves are afoot to harmonize the control of pesticides, and both countries have played a major role in this work.

*Japan.* In contrast to most countries involved in the harmonization of registration schemes, a few, such as Japan, are tending to act somewhat independently. Japanese regulations (Agricultural Chemicals Law; the Poisons and Deleterious Substances Control Law; and the Food Sanitation Law) now demand that certain toxicological and residue data which are required for registration must be obtained in official Japanese or other recognized laboratories, regardless of whether the data are available from other internationally accepted sources. The regulatory scheme for agricultural chemicals in Japan is shown in Fig. 19.4.

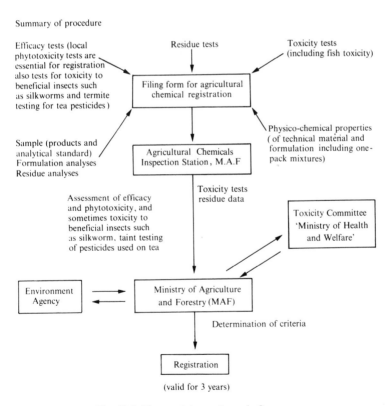

Summary of procedure

Efficacy tests (local phytotoxicity tests are essential for registration also tests for toxicity to beneficial insects such as silkworms and termite testing for tea pesticides)

Residue tests

Toxicity tests (including fish toxicity)

Filing form for agricultural chemical registration

Sample (products and analytical standard) Formulation analyses Residue analyses

Agricultural Chemicals Inspection Station, M.A.F

Physico-chemical properties (of technical material and formulation including one-pack mixtures)

Assessment of efficacy and phytotoxicity, and sometimes toxicity to beneficial insects such as silkworm, taint testing of pesticides used on tea

Toxicity tests residue data

Toxicity Committee 'Ministry of Health and Welfare'

Environment Agency

Ministry of Agriculture and Forestry (MAF)

Determination of criteria

Registration

(valid for 3 years)

Fig. 19.4. The regulatory scheme in Japan.

*United States.* In the United States considerable changes and amend-
ments are being made to the laws and regulations, and these appear
from time to time in the *Federal Register.* Useful summaries are to be
found in the issues dated 25 June 1975 and 3 July 1975.

The Federal regulation of pesticides began in 1910 and was
substantially expanded in 1947 and again in 1972. The second of these
expansions resulted in the complete restructuring of the Federal
Pesticide regulatory scheme so that the Federal Insecticide, Fungicide
and Rodenticide Act (FIFRA) changed from a mainly labelling law into
a comprehensive regulatory statute which controls the manufacture and
distribution as well as the actual use of pesticides.

This control of use of pesticides was a major change in outlook and
arose from the recommendations of the House of Representatives
Committee on Agriculture. The effect has been the regulatory control
of use and user, the banning of pesticides considered undesirable, and
the general streamlining and strengthening of enforcement procedures
to protect man and the environment against the misuse of these
biologically active materials. However, the committee did appreciate
the need for continued research for new pesticidal materials and methods
of pest control.

It is clear that Congress's primary purpose in enacting the Federal
Environmental Protection Control Act (FEPCA) was to ensure that
pesticide use was subject to a thorough review of environmental and
human health hazards. In keeping with this environmental and human
health perspective, the amended FIFRA has established many new
requirements of review in connection with the registration of pesticides.

*            *            *

This chapter has attempted to provide some information on the very
large amount of activity surrounding the legal and regulatory environment of pesticides. One hopes that it will have assured the reader that
both the pesticide industry and government regulatory authorities are
aware of their responsibilities for the safe use of products and that they
are implemented. Legislation in any sphere may be good or bad; but
there is no question that some form of control of pesticide usage must
be exercised, through regulations, to ensure that the products are
properly employed for the ultimate benefit of man and his environment.

### Further reading

1  BATES, J.A.R. Reflection on regulations – certain aspects in the official

control of pesticides in various countries. *Chemy Ind.* **103,** (20), 1324-31 (1968).

2 BESEMER, A.F.H. The pattern of legislation – the European pattern. *Proc. 3rd British Pest Control Conference (1971).*

3 COUNCIL OF EUROPE. *Agricultural pesticides,* 3rd edn. Strasbourg (1973).

4 FAO/WHO *Guidelines for legislation concerning the registration for sale and marketing of pesticides.* PL: CP/21, OH/69,3 (1969).

5 FAO. *A model scheme for the establishment of national organizations for the official control of pesticides.* ACP: CP28. FAO Working Party of Experts on the Official Control of Pesticides (1970).

6 FAO/WHO. *Codex Alimentarius Commission – Procedural manual.* 2nd edn. Joint FAO/WHO Food Standards Programme (1969).

7 FAO. *Ad hoc government consultation on pesticides in Agriculture and Public Health – Need for international standardization of basic pesticide registration requirements. Testing and environmental rules and procedure.* AGP: PEST/PH/75/B51 (1975).

8 GIFAP *Registration Technical Memorandum* No. 3. Groupement International des Associations Nationales de Fabricants de Pesticides, Brussels (1970).

9 GLASSER, R.F. Herbicides in perspective. The respective roles of government and industry – Part II. *Proc. 11th British Weed Control Conference* (1972).

10 ——, Industry viewpoint on legislation. *FAO/ICP Seminar on the effective and safe use of pesticides in agriculture and public health in Africa.* Nairobi, Kenya, October-November (1974).

11 ——, Growing dimensions in legislative procedures – labelling, residues and toxicology. *FAO/ICP Seminar section on the role of legislation at the national level in ensuring the safe and effective use of agricultural pesticides.* Costa Rica, April (1972).

12 ——, *Pesticides – International law and regulations. Paper presented at the* National Agricultural Aviation Conference, Las Vegas, December 1972.

13 PAPWORTH, D.S. Registration procedures and fail-safe precaution – towards an international standardization of pesticides data and evaluation procedures. *Proc. 4th British Pest Control Conference* (1975).

14 HAHN, S. Regulatory Schemes of Current Members of the EEC. *Proc. 11th British weed Control Conference* (1972).

15 SCHUHMANN, G. Pesticides regulation: safeguarding human health. *Span,* **16,** (2), 59-61 (1973).

16 TSCHIRLEY, F.H. Herbicides in perspective – the respective roles of government and industry – Part I. *Proc. 11th British Weed Control Conference* (1972).

17 WHO Control of Pesticides – A Survey of Existing Legislation. *Int. Dig. Hlth. Legisl.* **20,** 579-726 (1969).

Plate 20.1. *Left*, aphid (*Metopolophium dirhodum* Walk) of various ages, including a winged adult. A related species, *Myzus persicae*, can carry sugar beet yellows virus into a beet field and initiate a spread of disease that causes heavy losses. This can be done by a very small number of aphids and insecticides must be used to destroy them. *Right*, ladybird or coccinellid beetle (*Coccinella septempunctata* L.), which feeds on aphids. The insect, which has a scarlet back with seven black spots, needs to eat many aphids daily; it will fly away if the aphids are few and it cannot, therefore, entirely prevent or clear up an attack. The tiny aphid in the foreground is a newly born one. (Note: the adult aphid is about 4mm long, including the projecting wings, and the ladybird about 6.5mm.) (ICI photographs)

# 20 Alternatives to chemical pesticides

by D.L. Gunn

Before the present array of organic pesticides began to come on to the market about the time of the Second World War, protection against pest attack was seldom effective enough. The effective pesticides themselves readily damaged plants and animals.[8] Attention was therefore paid to control by non-chemical methods, which are nowadays presented as components of integrated control.

Use of modern chemical pesticides has much improved control; but because of increasing occurrence of resistance to pesticides and because of objections — often misplaced — raised to the use of synthetic chemicals, research workers are now concentrating more on integrated control, in which pesticides are used in conjunction with other methods.

In nature, uninfluenced by man, most species of animals and plants are attacked by diseases and by several species of predators and parasites. Even when food is abundant, a species population may thus be limited despite a reproductive capacity to produce large and rapid increase. Percentage losses of population often tend to be greater when the population is denser. Such natural control of numbers may be insufficient when dense and extensive crops must be grown in monoculture with little labour and crop losses must be minimized to feed townspeople who produce no food.

## Biological control

### Arthropods by arthropods

The control of arthropods by arthropods is man's calculated use of insect or mite predators or parasites to control the numbers of insect or mite pests. The greatest successes have been achieved in two circumstances: on a crop introduced from afar, with a pest but without its controlling agent, which is then introduced; and on oceanic islands with few species of animals and plants (even large islands like Australia and Hawaii) and ecological islands like California, shut off by deserts, mountains, and the sea. Successes have nevertheless been achieved in other circumstances, several of them recently.

The cottony cushion scale (*Icerya purchasi*) was accidentally introduced into California in 1868 on acacia plants from Australia. It spread to citrus trees so successfully that by 1886 ruin faced the citrus farmers. An entomologist sent to Australia to discover why the scale gave no trouble there returned with ladybirds *(Novius (Rhodalia= Vedalia) cardinalis)*. These produced quite satisfactory control, needing no more effort or expenditure at all. This established in California an esteem for biological control that has never since been lost. [4]

About the same time, the European gypsy moth *(Porthetria dispar)* was introduced carelessly into the eastern USA and quickly established itself on forest trees. By 1923 some 75 millions of insects of 45 species had been introduced to control the pest biologically, without success. It was not until DDT sprayed from aircraft was introduced that the spreading damage was checked. The use of DDT has now been forbidden. The forest destruction has begun again and is spreading.

Partial but inadequate success was achieved in controlling American blight or woolly aphis *(Schizoneura lanigera = Eriosoma lanigerum)* on apple trees in Britain by introducing the chalcid *Aphelinus mali.* By protecting the parasites over the winter, some control could be achieved; but too many woolly aphis survived and had to be controlled with insecticides, which would have done the job just as well without the aid of *Aphelinus.* In any case, biological control of only one of the many apple pests could hardly survive the intense spraying programme required to produce the high proportion of uninfested and undamaged apples demanded by the consumer. A comprehensive programme of biological control of all the pests of apple trees would be required and this is not yet in sight, though it is being sought.

T.H.C. Taylor became well known for his successful introductions of controlling insects into the island of Fiji in the 1930s. After many years' work on biological control, in 1955 he concluded that he knew it to be the best of all methods of controlling pests when it worked, but that it seldom worked and that there was little future for it in continental areas. [14] Taylor considered that in not more than 20 cases in the previous 50 years had control been so complete that nothing more was needed.

On the other hand, ten years later DeBach, [5] Professor of Biological Control at Riverside, California, claimed better results and prospects, with 66 complete successes (74 were reported by the Commonwealth Institute of Biological Control in 1969), 88 substantial successes, and 71 partial successes claimed: but nothing was said about the number of complete failures. There may be cases in which a partial success is useful, but commonly it leaves the insecticide requirements unchanged. For comparison with these numbers, the number of species of insect pests in the world is estimated at 10 000. Not all of these are important.

The outstanding success in Britain is the control system developed by Hussey for mites and insects on cucumbers grown in glasshouses, effectively a novel environment. [7] This system requires reliable supplies not only of the predator of the red spider-mite pest and the parasite of the white-fly pest, but also of the pests themselves. Similarly, the control of houseflies on a chicken farm, developed by the United States Department of Agriculture (USDA), requires supplies of the parasite *Spalangia endius* reared for releasing very much larger numbers than occur naturally. Only thus can the fly population be usefully reduced. The trade in rearing controlling insects for inundation releases may thus develop.

A formidable body of literature has been published on biological control, much of it from the enthusiastic group at Riverside. [4] Their comprehensive volume on the subject also attacks the use of insecticides with equal enthusiasm. In fact, a great acreage of use of pesticides seems to be entirely successful and harmless, skilfully managed, and ecologically sound. In the exceptional case, when the application of an insecticide does cause trouble, the bad news reaches the headlines and creates an impression out of proportion to the truth.

Research on biological control is now a big business but it is non-profit making even when it is successful. The work is done at non-commercial institutions such as the Commonwealth Institute of Biological Control, USDA centres and other governmental organizations, at certain universities., and at co-operative commodity research stations. The public is the beneficiary if produce is thereby improved or cheapened. On present knowledge most pests could not conceivably be controlled in this way but there still seem to be some openings for the method.

## Weeds

Weeds can reduce crop yields by competing with crop for light, water, and nutrients. They can also impede mechanical harvesting and their seeds may contaminate the produce.

In industrial countries today, labour is too costly for hand weeding. Chemical weedkillers are expected to kill most species of weeds and not damage the crop.[11] On the other hand, biological control agents have

to be chosen which attack few species. Consequently, chemicals are
generally used on crops, whereas biological control can be used where a
single species of weed has become dominant and troublesome, either in
a crop because of inherent resistance to the herbicides used or on
marginal grazing land of too little value to bear the cost of chemicals.
An example of resistant crop weed is *Chondrilla juncea* in South
Australia and Victoria, where an introduced rust, *Puccinia chondrillina,*
seems to be controlling it successfully.

The best-known case of biological control of weeds is the control of
several species of prickly pear *(Opuntia)* which, originating in the
southern U.S.A., had spread over 40 million hectares of grazing land in
Australia by 1925. Much of this land became unusable and in some
places the cactus was impenetrable. Imported enemies of the prickly
pear, amounting to 0.5 million insects of about 50 species, helped little.
Quite unexpectedly, a moth *(Cactoblastis cactorum)* from the
Argentine cleared out the cactus so completely that by 1930 nothing
else was required.

A.P.Dodd, who sent the *Cactoblastis* for trial, afterwards wrote:
'certainly its remarkable achievements could not have been foretold'.
This illustrates an important truth. When one intrudes into a situation
in nature, whether by growing a crop or by using a new insecticide or
by introducing a biological control agent, the results cannot be entirely
foretold. Care and research can help forecasting but one can never be
entirely sure, for real situations are far too complicated for forecasting
novel developments in complete detail. This is not, of course, a reason
for refusing to introduce novel techniques but it is a reason for taking
precautions to observe and check and to be constantly on the watch for
the unexpected.

The main danger in introducing an exotic plant parasite is that it will
not confine itself to the target species of plant but will attack crop
species and become a pest itself. Fear of such an attack delayed from
1922 to 1946 the introduction into northern California from
Australia of the European beetle, *Chrysolina quadrigemina,* to control
*Hypericum perforatum* (common St John's wort or Klamath weed).
The beetles cleared 800 000 ha of range land for grazing again. All such
introductions have to undergo a rigorous series of tests to exclude
possible pests.

*Microbes*

'Microbe' is an imprecise term, but it is convenient for including several small organisms parasitic on insects: bacteria, fungi, protozoans, rickettsias, and even eelworms (nematodes). [2] Many kinds of microbes may contribute to natural mortality in insects. The intention has been to infect some members of a population, hoping that the infection would spread naturally. So far this has seldom been achieved.

Early attempts were made by Metchnikoff in Russia in 1879 to use a fungus *(Metarrhizium anisopliae)* against cockchafers; by Snow in Kansas in the U.S.A. from 1890 to use another fungus *(Beauveria bassiana)* against chinch bug; and by D'Herelle in Central America about 1910 to use a bacterium *(Coccobacillus acridiorum)* to control locusts. [15] In each case it turned out that the insects died of the disease only in suitable weather; but then they did so anyway, whether the parasites had been added by man or not.

*Bacillus popilliae* has been known for a long time to control Japanese beetle *(Popillia japonica),* which attacks grass roots, by producing milky disease. The bacteria are used in the eastern U.S.A. when the high cost is considered worth while, as on golf courses and lawns, but because the pest is uncontrolled in rough grass, it continues to spread.

*Bacillus thuringiensis* has been much studied and is produced commercially for use in the field, but instead of spreading a disease it kills by toxic crystals within the organism. It is therefore better regarded as a complex insecticide produced commercially by a biological process.

Microbes that attack arthropods but which do not normally attack mammals are not necessarily innocuous. Allergic reactions occur at times. Both *Beauveria* and *Metarrhizium* have caused distress in vertebrates, though only rarely. Specific tests on mammals are required before microbial control agents can be registered. In any case, users should be warned to avoid heavy contamination (especially of their lungs and gut) by spores.

### Virus diseases of arthropods

Many known viruses attack insects. These are: nuclear polyhedrosis viruses (NPV), the best known group, 170 in number; granulosis viruses (GV) (50); and 136 other kinds, including the iridescent viruses of *Tipula* and mosquitoes and nine pox viruses. There is also another kind of virus potentially important for controlling red mite. The NPV and GV are known only in insects and are unlikely to harm organisms in other classes, but some of the other classes of viruses contain some that cause serious diseases in sheep, fishes, plants, and fungi, while certain pox viruses cause fatal disease in man. [11] Only NPV and GV have been much tested against insects. [2]

NPV have the valuable property of persistence for years, within a polyhedral coat, provided they are kept dry or in water in a cold room and are not exposed to sunshine. Iridescent viruses do not survive long outside a host except in a deep freeze. Propagation of viruses always requires to be done inside living cells. Transmission is commonly on food, through the mouth, or the virus may be picked up from the eggshell by the hatching larva.

Insect populations sometimes harbour a virus for generations without showing disease symptoms until they are put under stress, for example by overcrowding or by sub-lethal doses of insecticide. Then the disease appears. One result of the occurrence of such an occult infection or balanced parasitism is that the disease may intervene as a density-dependent check on population but, by the time it acts, damage will have been done. For example, in Oregon, U.S.A., the tussock moth *(Hemerocampa (Orygia) pseudotsugata)* attacks Douglas fir and has been controlled by DDT sprayed from aircraft. When the use of DDT was forbidden by the United States Environmental Protection Agency, an expected outbreak of virus disease was relied upon to occur when the moth population became dense. But it did not come and in 1974 the Agency eventually yielded to the outcry against unnecessarily permitting immense and long-lasting damage to the forests.

Viruses are most useful in control where some damage can be tolerated and where the value of the crop will not bear the cost of the required use of insecticides. If the virus is abundantly sprayed on the crop early in the pest's life instead of being transmitted naturally, damage can be reduced. Early application is important for pests that burrow and cannot then be reached. Some cotton pests have developed resistance to several pesticides and early virus application might be particularly useful (see Chapters 13 and 16).

Cabbage butterflies, *Pieris brassicae,* caught in several European countries exhibit various degrees of susceptibility to their specific GV and a laboratory stock of some years standing that had been through a GV epidemic was even less susceptible.

On the whole, insect viruses seem to be fairly specific. This is advantageous in that their attack is confined to target species, but disadvantageous in that cost of development and testing for registration (millions of pounds sterling) could seldom be recouped in controlling a single species.

One wide-spectrum virus attacks not only several species of *Heliothis* larvae but also other genera, including even some predators. It could control cotton bollworm *(Heliothis (Helicoverpa) armigera)* and also tobacco budworm *(Helicoverpa virescens),* and corn earworm *(Heliothis zea)* which sometimes attack cotton. The virus can infect cabbage looper, beet armyworm and pink bollworm, some of which are destructive in fields of lettuce. [9] In 1973 in the U.S.A. it was exempted from tolerance when used on cotton, as not posing a health hazard and was permitted for experimental use as Viron H and Biotrol VHZ. It is, however, unstable above 38°C and is therefore short-lived in tropical sunshine. In October 1975 a variant of Viron H named Elkar had been registered by the United States Environment Protection Agency for use in the U.S.A.; the efficacy was consistent with what the label claimed.

The most valuable introduction of a virus occurred accidentally. Insect parasites were taken from Europe to Canada to help control European spruce sawfly *(Neodiprion sertifer).* Apparently they carried a virus which controlled the sawfly dramatically. It has since been deliberately spread to other parts of Canada and the U.S.A. and is now considered to be the main controlling agent. No sign of resistance has appeared.

## *Other biological control*

A familiar example outside the arthropods is the introduction of a virus disease, myxomatosis, from South America to control rabbits in Australia. This failed several times before it succeeded. From time to time and place to place, rabbits increase locally but then the disease hits them again and numbers are reduced by it. This is characteristic of successful biological control and occurs also with prickly pear.

A less fortunate example is the introduction into the island of Jamaica of the mongoose *(Herpestes auropunctatus)* to control black rats *(Rattus rattus).* This was successful until the rats changed their habits and evaded the mongooses. Then the mongooses became a pest by attacking domestic poultry. [16]

Control of a plant virus disease by another virus is now being used. In tomatoes a severe disease is caused by a virulent strain of tobacco mosaic virus; the disease is prevented or much reduced by previously infecting the plants with a closely related non-virulent strain of the virus.

247

## Genetic control

### Breeding plants for resistance

The main purpose of most plant breeding programmes is to increase crop yields.[11] This is often achieved by cross-breeding or by inducing mutations and selecting more productive lines. Increased yields can come from plants that are more responsive to fertilizer, those that have shorter stems and therefore stand up better to adverse weather, or those that resist attack by pests and diseases.

The best known breeding programmes for resistance are those of the U.S.A. and Canada which produce varieties of wheat able to resist attacks of black stem-rust, *Puccinia graminis tritici*. Unfortunately, the stem-rust also produces its own varieties, which eventually overcome the resistance bred into the wheat, and these new lines quickly become the dominant varieties because they are the only ones that can propagate on the resistant wheat. Wheat breeders have to produce new varieties frequently to meet this recurring challenge.

This is but one of the many long-established programmes for breeding fungus-resistant varieties.[8] Resistance to insect attack is only rarely achieved but a variety of the cotton plant *(Gossypium)* resistant to attack by leafhoppers *(Empoasca facilis)* has been produced. Only this variety (U4) can be grown profitably in South Africa. In other cases, resistance to insect attack varies with changes in environmental conditions or may be defeated by a change in the pest.[5] Painter wrote a comprehensive and authoritative review of the subject in 1951.[10]

Similar to breeding is the use of a pre-existing resistant variety. Thus, the European wine industry was saved from ruin a century ago by the practice of grafting their own non-resistant but high-quality vines on to American root stocks resistant to the attacking aphis, *Phylloxera vastatrix*.

Strong resistance to pest attack is now required by some animals. Australian beef cattle are attacked and damaged by nine species of ticks which — like spider-mites — develop pesticide resistance remarkably quickly. Although, with most species of plants and animals, strains could probably be produced that would be resistant to given species of insects, the demands made on the breeding institutions are so many, especially to increase yields, that the limited resources available tend to be devoted to needs for which breeding provides the only solution.[11] In most cases, however, insects can still be controlled by other means.

There is another difficulty. Genes for resistance to given insect species have to be combined into a variety having already high potential yield and resistance to diseases against which there is no other feasible defence. This process tends to take years, though it is quickened by new methods; but by the time it is complete, the disease resistance of the variety may have been broken and the variety replaced in agricultural practice by a different one. This kind of risk can be weighed up only by expert breeders who are also familiar with commercial production – and then they may turn out to be wrong.

*Insect pests*

Davidson has comprehensively described the methods proposed for genetic control of insect pests.[3] They are applied to one generation to reduce the population in the next generation. Most of the methods proposed have not yet been tried, or only on a small scale. They involve chromosomal or cytoplasmic incompatibility or deliberately produced chromosome translocations. Releases of fertile insects that are infertile with the wild population must be restricted to one sex (in mosquitoes, the male) lest a new pest variety be introduced to the locality. The task of separating many thousands of mosquitoes each day into the sexes *without any mistake* is a daunting one.

The only well-known method is the release of great numbers of the pest species that have been sterilized by gamma rays or chemicals, so that they mate but produce no progeny and tend to exhaust the reproductive capacity of their fertile wild mates. The now classical control of screw-worm *(Cochliomyia hominivorax),* attacking cattle in the southern U.S.A., entailed the rearing, sterilization by gamma radiation, and release from aircraft of 150 million flies per day at the height of the action. This required 100 tonnes of horsemeat and 10 tonnes of dried blood each week. The dose of gamma rays had to be controlled with care, so as to sterilize all the flies but not to impair the vigour of the males in searching for and copulating with wild females, nor to reduce their longevity.

After a series of trials starting about 1955 in ecologically isolated areas of Curaçao and Florida, the main operations covered the cattle areas of Louisiana, Oklahoma, Arkansas, Arizona, Texas, and New Mexico. Despite the great cost, including that of extending releases deep into Mexico, because flies immigrated thence, and failure to eradicate the species, for years the community profited from fewer cattle deaths, less damaged hides, and cattle in better condition. On oceanic islands, a species has occasionally been eradicated by this technique.

In 1972, for reasons that are still mysterious, control of screw-worm broke down and nearly the original amount of damage was soon occuring annually.

Knipling has published simplified calculations to show that this method of control is more effective than insecticides and can eliminate a species, which insecticides are unlikely to do. The calculation is, however, not generally valid because it takes no account of density-dependent checks (which tend to allow more survivors from natural hazards in reduced populations) and of rapid rates of reproduction (which, with some species, could result in total failure to reduce the population unless impracticably large numbers of sterilized insects were released). On the other hand, if the reproductive potential is small, the number released need not be so large, but then the rapid production of many insects for sterilization and release is difficult, as with tsetse flies. A full theoretical treatment is quoted by Davidson.[3]

Before undertaking the expense of a field trial, one must know the population size, preferably at its smallest. This population has to be greatly exceeded by the number of sterilized insects released. Results can be reasonably interpreted only if the population size is assessed repeatedly. With some species, such as the greenbottle blowfly in Britain, responsible for sheep myiasis, attempts to assess population size have so far failed.

Sterilization by feeding certain chemicals is also possible and avoids the expense of safely using large-scale radiation techniques. Unfortunately, many of the available sterilizing chemicals act by producing harmful mutations in mammals (including human beings) as well as in insects. Some are carcinogenic as well as mutagenic. They cannot, consequently, be freely exposed or used with bait, for they might get into human food or be eaten by pets or desirable wild animals. Nevertheless, a field trial was made on a refuse dump in Florida and it successfully controlled a fly nuisance.

Considering the hazards a species has survived through millenia, it is not surprising that resistance against sterilants develops just as it does against insecticides and viruses. Houseflies showed resistance to apholate after 15 generations. After 20 generations a tenfold dose was required, and after 35 generations, a 26-fold dose. Similar results with mosquitoes have been found with other sterilants.

**Hormones, hormone bioanalogues, and hormone antagonists**

Complex processes such as growth, moulting, pupation, and excretion in insects are controlled and co-ordinated by hormones being moved about the body. Their action can be imitated in some cases by foreign compounds which are not necessarily closely related in structure to the true hormones. They are thus not chemical analogues, but analogues only in the biological sense of producing similar results in the living organism. Applying such chemicals, natural or not, at the wrong time destroys proper co-ordination or antagonizes a hormone that should be working. These are, of course, kinds of chemical control but are usually distinguished from conventional insecticides.[13]

In the 1950s, the idea of using hormones for controlling insect pests was welcomed under the title 'third-generation insecticides'. But they have been slow in growing up. If juvenile hormone is applied at the end of larval life, when it should be disappearing, large larvae which are abnormal are produced; but the application affects only individuals at that sensitive stage. The hormone is not specific. An internal parasite, *Aphidius nigripes,* is more susceptible to it than its aphid host. In a normal, slow-growing population of mixed ages, application has to be made repeatedly, for the materials are not persistent. Populations of uniform age seldom occur. These materials are unlikely to be registered in the U.S.A. because of the huge cost of the tests required by the registering authority.

Moulting hormones are steroids, a group containing several mammalian hormones, including sex hormones important in human beings. Steroids are costly to synthesize and extraction from insects is not practicable for field use.

At first it seemed likely that an insect species could not develop resistance to its own hormones; but the detoxication mechanism known to exist, which inactivates the hormone in normal life, can be strengthened and is indeed known to be more active when the hormone is untimely; and cuticle can become less easily penetrated. If a bioanalogue is used, a detoxication system could emerge selectively eliminating it without affecting the natural hormone.

The future of control by hormones is not rosy. There is perhaps more promise in studies of hormone antagonists, and also of derivatives of juvenile hormones and their bioanalogues, used simply as new synthetic insecticides or female sterilants. A few are now commercially available for use on mosquitoes and aphids.

251

## Pheromones

Pheromones are chemicals or mixtures of chemicals secreted by animals, which influence the behaviour of other animals of the same species. They are particularly important in short-lived animals, which cannot accumulate experience and must react quickly. In insects they are most important as sex attractants, generally signalling to distant males that a female is present up wind. There are three possible ways of our using them: to bait traps for estimating numbers in a population; to catch so many males that there are not enough left to maintain the population; or to blanket the air with so much pheromone that the males become confused.

An exceedingly small amount of pheromone may be effective. Thus, an extract from virgin female pine sawflies *(Diprion similis)* amounting to less than $10^{-6}$ g, attracted 500 to 1000 males within 5 minutes. The natural chemicals have been identified in several cases but in practice synthetic chemicals must be used.

Fruit-fly and melon-fly populations have been enumerated using traps baited with pheromones. In north-east U.S.A. traps baited with 'Dispalure' have been used to detect the advance of gypsy moth. Other commercially available materials are 'Muscalure' (flies), 'Codlemone' (codling moth), and 'Grandlure' (boll weevil).

Mass trapping has succeeded experimentally with small populations but failed with heavy infestations. The blanketing or confusion technique is being tested in the U.S.A. In a dense population, however, males could find females simply by just searching, without any pheromone aid.

Thus, apart from the use of pheromones in population assessment traps, for which food and light can be used in some species, the value of pheromones in large-scale control programmes remains to be demonstrated.

## Integrated control

Before pesticides could be used in field practice (except in high-value situations like orchards) because of their high costs and their side effects, simple traditional methods were not without value. Rotation of crops, partly to check pests specific to one crop; fallowing for a year; using widely spaced rows suitable for weeding by hoe or machine instead of chemicals; choosing the date of planting so that the plants would not be at a sensitive age when the annual flush of their pests occurred; selecting methods of cultivation and other components of good husbandry; all these methods contributed to the reduction of weeds, pests, and diseases. Such cultural methods combined with, and modified by, modern chemical techniques are today fashionably called 'pest management' or 'integrated control' — terms which embrace any combination of artificial and natural controls to produce the greatest advantage.

Integrated control may include changing methods of husbandry or variety of crop, and taking biological advice to indicate whether and when pesticides ought to be used in relation to potential damage. Benefit to the producer is the primary financial aim. This is largely dependent on the rate of crop yield, but there are also longer-term considerations. Any farmer who ignored his financial benefit in favour of consideration for wild plants and animals would soon have to drop out of farming, unless he had an income from elsewhere. The greatest number of farmers in Britain hold no more than 33 ha and get a net income often less than the wage in many safe and routine manual jobs requiring no capital or risk. Thus, to be adopted, a change of system must pay.

Most good examples of integrated control are older than the phrase (1959).[7] In Britain, sugar beet is protected by a combination of methods required by the only purchaser of the product, the British Sugar Corporation.[11] Virus yellows disease of beet is transmitted by an aphid, *Myzus persicae,* from plants that survive the winter. Beet produces seed only in its second year from sowing; so seed beet is planted far away from that used for sugar, because the aphids can over-winter on the seed crop. Proximity to mangold clamps has to be avoided. If the routine field surveys by trained men reveal more than one aphid per four plants, a density that can result in 20 per cent loss of yield, a systemic insecticide is applied. This is demeton-methyl, which enters the plant and kills the aphids that suck the sap. This only restricts the spread of the disease to some extent. It is effective when backed by hygienic measures, so that infection pressure is low.

Coccinellid beetles (ladybirds) eat many aphids in a dense population but are useless against those sparse populations that can infect the whole crop with virus yellows and cause much loss.

Infection by cyst nematodes, pygmy beetles, and other pests has been minimized in Britain because the British Sugar Corporation requires all planting to be done on land on which no beet or brassica crop has been grown in the previous 2 years; a 4-year interval is actually common.

Some of the best examples of established integrated control are practised on large plantations in Africa, Asia and the U.S.A. In most developed countries, on the other hand, vegetable and fruit crops are grown on a smaller scale for individual human consumption or processing and the market demands a very high standard of freedom from pests and pest damage. Biological control may only seldom have something to contribute in these circumstances but other components of integrated control may substantially help — and indeed usually have done — to reduce the need to use pesticides; nevertheless, pesticides seem likely to remain the main weapon of the producer in the foreseeable future.

*      *      *

Most of the reviews of any alternative to pesticide control are written by specialists in that alternative who show a proper enthusiasm and optimism about the future; but other people must assess the situation more coldly, so that development will be realistic.

Propaganda against pesticides may weaken as the consequences of widespread food shortages become obvious;[6] but pest resistance to pesticides is a different matter. It will not just go away and is likely to continue to become commoner: the first requirement is to make use of what we know to delay its increase (see Chapter 16). Some general way of defeating resistance, not yet conceived, may be developed. Otherwise man must constantly discover new varied devices to control pests. The methods outlined in this chapter help in a few important cases but still fail to match insecticides in generality.

## References

1    ANON. The use of viruses for the control of insect pests and disease vectors. *Wld. Hlth. Org. Rep. Ser.* **531**, Geneva (1973).

2    BURGES, H.D. and HUSSEY, N.W. (eds.) *Microbial control of insects and mites.* Academic Press, London (1971).

3    DAVIDSON, G. *Genetic control of insect pests.* Academic Press, London, (1974).

4    DEBACH, P. and SCHLINGER, E.J. (eds.). *Biological control of insect pests and weeds.* Chapman and Hall, London (1964).

5    —— , *Biological control by natural enemies.* Cambridge University Press (1974).

6    FLETCHER, W.W. *The pest war.* Blackwell, Oxford (1974).

7    HUFFAKER, C.B. (Ed.) Biological control. *Proc. AAAS Symp. Boston, December 1969.* Plenum Press, New York (1971).

8    MARTIN, H. *Scientific principles of plant protection.* 1st edn. Arnold, London (1928); 6th edn. Arnold, London (1973).

9    MARX, J.L. Insect control, II: Hormones and viruses. *Science, N.Y.* **181**, 833-5 (1973).

10   PAINTER, R.H. *Insect resistance in crop plants.* Macmillan, New York (1951).

11  PRICE JONES, D. and SOLOMON, M.E. (eds). Biology in pest and disease control. *13th Symp. Brit. Ecol. Soc., Oxford, January 1972.* Blackwell, Oxford (1974). (See especially contributions by Cussans, Gunn, Hull, Lupton, and Tinsley & Entwistle).
12  RIVERS, C.F. Virus resistance in larvae of *Pieris brassicae. 1st Int. Conf. Insect Path. Biol. Control, Prague, 1958.* 205–10 (1958).
13  SLAMA, K., ROMANUK, M., and SORM, F. *Insect hormones and bio-analogues.* Springer-Verlag, Vienna and New York (1974).
14  TAYLOR, T.H.C. Biological control of insect pests. *Ann. appl. Biol.* **42,** 190-96, (1955).
15  UVAROV, B.P. *Locusts and grasshoppers, a handbook for their study and control.* Imp. Bur. Ent., London (1928).
16  YAPP, W.B. *Production, pollution, protection.* Wykeham Publications, London (1972).

## Further reading

1  EMDEN, H.F. VAN *Pest control and its ecology.* Institute of Biology's Studies in Biology, no 50. Arnold, London (1974).
2  FLETCHER, J.T. and ROWE, J.M. Observations and experiments on the use of an avirulent strain of tobacco-mosaic virus as a means of controlling tomato mosaic. *Ann. appl. Biol.* **81,** 171-9 (1975).
3  GUNN, D.L. Dilemmas in conservation for applied biologists. *Ann. appl. Biol.* **72,** 105-27 (1972).
4  MELLANBY, K. *Pesticides and pollution.* Collins, London (1967).
5  SOLOMON, M.E. *Population dynamics.* Institute of Biology's Studies in Biology, no. 18. Arnold, London (1969).
6  WOOD, R.K.S. (ed.) *Biological problems arising from the control of pests and diseases.* Symposium no. 9, Institute of Biology, London (1960).

# Appendix: Pesticides: a guide to terminology

By I.D. Farquharson

## What is a pesticide?

The idea that all materials used to control pests do so by killing them is too facile to encompass the full array of sophisticated products that have been developed over the years with the aim of improving the productivity of agriculture and of protecting man, animals, and the environment from pest attack.

To see the picture clearly, one first has to ask the question, 'what is a pest?' and one is soon led to the conclusion that there is no absolute definition which will be universally acceptable. An organism – plant, animals, or microbe – which may in itself be damaging in that it is competing with man in the economic or aesthetic use or enjoyment of his environment, is not a pest until it does so at a level which is intolerable. What is intolerable to one person may be tolerable to another: the fruit grower will be in conflict with the amateur ornithologist in his attitude to the bullfinch; the poultry farmer with the naturalist in his attitude to the fox. Moreover, an organism which may be regarded as a pest in one geographical region or by one ethnic group may not be so classed by another: a Masai warrior-herdsman is used to flies alighting on his face, while a European is likely to be irritated by a single insect. Such conflicts and differences inevitably have their repercussions in attitudes to pest control.

Perhaps there is no better way of describing a pest than that used by Mellanby [1] when he says that it is a plant or animal living where man does not want it to live. There are a variety of reasons why man may not want certain plants or animals – or, indeed, viruses and bacteria, which are also pests: for instance, because the organism could cause economically or aesthetically unacceptable damage or loss to crops, livestock, or produce in storage, or to personal or public property; or because it encroaches on land otherwise suitable for agriculture, industry, or leisure activities; or because it poses a threat to human, animal or plant health as a vector of diseases; or because in some other way it prevents man's use or enjoyment of an otherwise agreeable environment.

The definition of a pesticide is no easier. Certainly in modern usage it is no longer confined to the killing of pests, as is implied by the derivation of the term. 'Pesticides' may include bacteria and viruses for the control of insect pests, chemicals which inhibit the normal growth and reproduction of pests, or which so affect their normal behaviour patterns (e.g. mating, feeding or swarming) that they no longer pose a threat as pests. It is perhaps more accurate, therefore, to talk of pest-control agents and to describe them as preparations of synthetic or naturally occurring chemicals or pathogens used to reduce pest populations to an acceptable level or to prevent damage which they cause. The definition excludes physical agents such as electric grids, sterilization by irradiation, the use of non-microbial predators and parasites, and the breeding of genetically resistant varieties of plants and animals.

Even this definition is not sufficiently broad to include an emerging group of chemicals derived from research in pesticides and usually classed with them. These are the 'plant growth regulators' (PGRs), which are a logical extension of work on weed control aimed at disrupting the normal growth of plants. They are chemicals which are not normal plant nutrients in the sense of fertilizers and trace elements, but which affect the growth of plants in a beneficial way, for example by improving the yield, quality, or climatic tolerance of crops or assisting their harvesting. For this reason it is becoming fashionable to talk of 'agrochemicals', although even this title may be confusing in that many have non-agricultural uses, while the term may or may not be used to include fertilizers and plant nutrients, food additives, and agricultural uses of other chemicals such as plastics.

## Main uses of pesticides

The primary classification of pesticides is in terms of their use, i.e., the group of organisms controlled, although frequently more than one term may be in common use for a particular group and individual products may have a range of applications that crosses the boundaries of such a classification.

*Insecticides* are products designed primarily to control insect pests. They can be subdivided, according to the main field of use, into crop insecticides; veterinary insecticides (ectoparasiticides) to control external insect parasites of animals; public health insecticides to control insect vectors of human diseases; stored-product insecticides; and household (domestic) insecticides. Crop insecticides are, in addition, said to be foliage or soil insecticides according to whether they are applied to the leaves or to the soil to protect the root system. Another important subdivision is between systemic and contact insecticides. This is discussed on page 263.

*Acaricides* (miticides and tickicides) are mainly used to control plant-feeding mites (crop acaricides) and mites and ticks which parasitize animals (veterinary acaricides). Mites and ticks belong to a zoological group known as the Acarina which are not strictly insects; but acaricides are often included in the general term insecticides, and insecticides which are active against mites and ticks are said to have 'acaricidal activity'. Products designed solely for mite control are known as specific acaricides.

257

*Fungicides* are products which control fungal diseases of crops, stored produce, and fabrics. (Medical and veterinary fungicides are generally classed as pharmaceutical chemicals). Crop fungicides are further sub-divided into foliage fungicides, which are applied to the emerged parts of plants; soil fungicides, applied to the soil to control soil-borne diseases; and seed dressings which are used to provide seeds with a protective coating (seed dressings may also incorporate insecticides, alone or with a fungicide). A further distinction is between protectant fungicides, which prevent infection, and eradicant fungicides which will overcome an existing infection.

*Bactericides and viricides* are often — erroneously — included in the use of the term 'fungicides'; they are products which control bacterial and viral diseases of crops, but in the context of pesticides they usually exclude industrial bactericides and disinfectants as well as anti-microbial drugs for animals and man.

*Herbicides* (weedkillers) are products used for the control of unwanted vegetation. They fall into two main catagories: those which are applied to the soil and are taken up by the plant through the roots: and those which are applied to the foliage and are taken up through the aerial parts of the plant. Herbicides which are taken up through the roots can also be used to prevent infestation by weeds by applying them to the soil before the weeds have emerged. Generally, a herbicide is tolerated more by some plants than by others — as, for example by grasses (classified botanically as monocotyledons) rather than by broad-leaved plants (which are dicotyledons). Where the tolerant plants are a crop, the herbicide is said to be selective for that crop. Different crops therefore generally require different herbicides, even where they may suffer from the same weeds. At high dosage rates, selectivity can disappear and the same herbicide may be used for non-selective weed control on industrial sites, railways, and elsewhere, where a complete clearance of vegetation is required. This type of application is referred tò as 'total weed control'. As there will always be some plants which will tolerate a particular herbicide, irrespective of dosage rate, the commonly used terms 'selective weedkiller' and 'total weedkiller' are not absolute, but refer to the way in which the herbicide is used. Herbicides are also classified as pre-emergence and post-emergence products, according to the stage of crop or weed growth during which they are normally applied (see also 'Mode of action', page 263).

*Nematicides* are products which control the microscopic organisms known as eelworms or nematodes, which parasitize plants. These pests usually attack the root system; hence nematicides are generally applied to the soil. One important group comprises the fumigant nematicides *(soil fumigants)*; others act primarily through contact or by penetrating the tissue of the plant (contact and systemic nematicides). Many nematicides have activity against other classes of organism. Those that are solely active against nematodes are known as specific nematicides.

258

*Soil fumigant* is a term often used, erroneously, as a synonym for nematicide because many of the nematicides in common use are soil fumigants. Correctly used, the term describes products which control any organism in the soil through activity as a vapour.

*Soil sterilants,* strictly defined, rid soil of all living organisms (i.e., fungi, bacteria, insect larvae, nematodes, weeds, and weed seeds) likely to damage plants that are subsequently grown in the soil. In practice, chemicals classed as soil sterilants may have incomplete or virtually no activity against one or more types of organism. Most are soil fumigants, in that they are active as a vapour; the term 'soil sterilant' implies activity against a broad range of organisms, including, particularly, microbial pathogens, whereas some soil fumigants (q.v.) are specific to nematodes. Soil sterilants may also be classed as soil fungicides or as nematicides by virtue of their prime activity or the main purpose for which they are used.

*Rodenticides* are products to control rodent mammals, in particular, rats, which can damage crops, property, or stored produce or can be a danger to public health as vectors of disease. Those in common use fall into two main classes: anti-coagulants, which prevent the normal blood-clotting mechanism of the animal; and stomach-acting rodenticides; both are generally used in baits. Space fumigants, which are active as a vapour, are also used to protect stored produce from rodents as well as from insect pests.

*Molluscicides* are used to control slugs and snails (molluscs) which feed on crops, or which are pests of veterinary or public health importance through being vectors of the parasites of animals and man known as flukes; these cause liver fluke disease in livestock and bilharziasis (schistosomiasis) in man. Products for the control of flukes themselves, with products used to control nematode and tapeworm parasites of animals, are referred to as *anthelmintics;* these are generally classed as drugs, but several were discovered during research carried out by the pesticide industry.

## Chemical classification of pesticides

The detailed classification of pesticides is based on chemical groups. Some have become very well known; the chlorinated hydrocarbon insecticides or organochlorines, for example, have achieved a certain notoriety because this class includes some of the insecticides which have received greatest criticism from the point of view of persistent residues in the environment. Today, there are so many compounds in use, and the complexity of chemical classification is so great, that it would not be practicable to discuss them here. The table at the end of this appendix sets out the main groups of pesticides and gives examples of some of the better-known compounds in each category.

But a word of caution is necessary. Valuable though classification may be as a convenient method of reducing a complex and diverse range of chemical terms to manageable proportions, there is danger in generalization, as just two examples show. First, as has been said above, organochlorines have achieved a certain notoriety in relation to the problem of persistent residues; yet many non-persistent pesticides which are not normally considered as organochlorines could rightly be classed as such, because they contain carbon, hydrogen, and chlorine; while some organochlorines (for instance, methoxychlor, which is closely related to DDT) have the attribute of short persistence. The second example is that of the organophosphate insecticides, a large group which is remarkably homogeneous chemically and in the way in which the different compounds act on animal systems. But members of this group exhibit a diverse range of physical and biological properties which defies any other generalization about their characteristics.

## The main features of pesticides

*Active ingredient.* 'Active ingredient' (or 'active principle') is the term applied to the chemical or chemicals in a product which are active pesticides. The active ingredient often forms only a small proportion of the product as sold to the farmer, the proportion usually being referred to as a percentage by weight (w/w) or by volume (w/v) of contained active ingredient.

The active ingredient may be referred to by its chemical name (for example, copper sulphate), but more commonly a 'common name' is assigned to highly complex molecules by international agreement. This is normally derived from the chemical name: dichlorvos, for example, is the accepted common name for the insecticidal compound 2,2-dichlorovinyl dimethyl phosphate.

Products containing one or more active ingredients may also be referred to by the trademark of the manufacturer or distributor. In literature this is distinguished from a common name by being given a capital initial letter and, in some cases, the symbol ® to indicate that it is a registered trademark. Often, there are several trademarks in existence for products containing the same active ingredient, according to the commercial origin of the material, and sometimes varying from country to country. Different trademarks may also be used to distinguish between products containing the same active ingredient but designed for different purposes: for instance, when the same active ingredient is used in products to control both crop and livestock pests.

*Formulation.* For a variety of reasons it is not usually possible to apply the active ingredient as such in the field. The first reason is that one cannot usually manufacture an absolutely pure chemical on a commercial scale. The chemical, as manufactured, is referred to as the 'technical material', with a defined minimum purity in percentage terms, as close as is technically and economically practicable to 100 per cent. Secondly, most pesticidal chemicals produce their effects when used in such minute amounts (for example, measured in grams of active ingredient per hectare of crop) that it would be quite impracticable to apply such a small quantity evenly over the crop. One of the most obvious ways of overcoming this would be to dilute the active ingredient with water to allow it to be applied in the form of a spray. Few chemicals, however, are sufficiently soluble or sufficiently stable in water to allow this. Instead, the technical material has to be combined with other ingredients to form a product which the farmer can dilute in the field. This is known as 'formulation'.

The two most common formulations for sprays are emulsifiable concentrates (ECs) and wettable powders (WPs). In EC formulations a solvent in which the active material will dissolve is used together with emulsifiers which will produce an emulsion when mixed with water. In WP formulations a finely divided inert powder acts as a 'carrier' for the active ingredient, and wetting and suspending agents are added to ensure quick and even dispersion when diluted with water.

Other materials are commonly used to enhance the performance of the final product. Thus, stabilizers may be added to prevent the active ingredient decomposing, and wetting agents can be added to improve the retention and spread of the spray droplets on the surface of the leaf. The ingredients of the formulation which are not biologically active are collectively known as inert ingredients, the active material content being referred to in terms of contained active ingredient (e.g., a 20 per cent EC).

There is, in addition, a special category of chemicals, which, although not themselves active, will increase the inherent activity of a particular active ingredient. These are known as synergists and they are said to display synergistic activity, or synergism. The same term is applied if two active ingredients used together have a greater activity than would be expected from the simple combination of their activity when used alone.

The two examples of formulations already mentioned are examples of spray formulations, ECs being liquid formulations and WPs solid formulations. Sprays are generally the most convenient and effective way of applying pesticides, but in certain circumstances (for example, where water is in short supply or for combined application with seed into the seed drill) it is more practicable to have a solid formulation which can be applied in the form of dusts or granules. In these, the active ingredient is incorporated into, or coated on, a solid carrier, such as a refined clay. Such formulations may be manufactured ready for use, but dusts, because of the cost of transporting large volumes of product, are often produced as a dust concentrate, which can be further diluted near the site of application by adding a locally available clay diluent or filler to form a field-strength dust.

These are the common formulations; others which have been developed for particular circumstances, for instance, where the product is particularly insoluble in water, are dispersible liquids, suspension concentrates, and oil formulations. In the few instances where the active ingredient is soluble in water, it may still be convenient to formulate it either by the addition of additives or by dilution in a water-soluble solvent or carrier to form soluble powders or water-soluble concentrates.

The design of equipment, as well as formulation, is an important aspect of the efficient use of pesticides, and often formulations and equipment designed for the particular purpose for which the pesticide is to be used, go hand in hand. Equipment can be designed for manual use (e.g., knapsack sprayers and granular or dust applicators), or for use with agricultural machinery (e.g., tractor-mounted or aerial spraying equipment). It can also be designed for very specific purposes: for example, seed-dressing machinery for coating seed with pesticides, for which special seed-dressing formulations are required.

For spray applications, according to circumstances and the type of equipment used, the required quantity of active ingredient may be applied to a given area by diluting the formulation in relatively small or large volumes of water or other diluent. 'High-volume' spraying is the application of thousands of litres of spray per hectare: 'Low-volume' spraying is the application of quantities in the region of tens to hundreds of litres of spray per hectare; while to apply very small quantities of spray evenly over wide areas in circumstances where weight is a critical economic factor (as in aerial spraying), special ultra-low-volume (ULV) formulations and equipment have been developed.

There are other reasons, besides the practicability of application, why active ingredients are sold as formulated products. One of the most important is that the properties of the active ingredient can often be modified by the formulation so that disadvantageous properties (such as high toxicity to man) can be minimized, and advantageous properties (for example, ability to get rapidly to the site of pest attack) can be used with optimum effect. An example is the insecticide dichlorvos, mentioned above. This has the disadvantages of being relatively toxic and so volatile that its effectiveness as an insecticide is of very short duration; but it also has two great advantages: it is highly active and it acts rapidly on certain insects of public health importance; and it has extemely short persistence in the environment. By formulating the active ingredient in a plastic material, from which the vapour is slowly released over a long period (an example of a slow-release formulation) it has found wide use as a domestic and public health insecticide.

*Mode of Action.* The way in which a pesticide acts is an important property which is frequently used in classification of different active ingredients. When the pesticide has to be eaten by the pest before it is effective it is said to exert 'stomach action'; when it acts through penetration of the skin (cuticle) it is said to have 'contact action'; and when it acts through inhalation of the vapour it has 'fumigant action'.

Mode of action, in the more basic sense of how the pesticide acts on the physiological system of the pest, has also introduced a range of scientific terms which are used to distinguish between pesticides. Two of these terms are in particularly common use. First, 'hormone weedkiller' describes herbicides (specifically the phenoxy acids, such as 2,4-D and MCPA) which, like naturally occurring plant hormones, stimulate growth, but do so in an abnormal way, so distorting the normal growth pattern of broad-leaf weeds that they become weak and eventually die. Second, 'cholinesterase inhibitors' is used to describe a category of insecticides (covering most products belonging to two of the main chemical groups — the organophosphates and the carbamates) which rely for their activity on inhibiting or competing for acetyl-cholinesterase, an enzyme important in the transmission of nerve impulses.

Another classification which is used, particularly referring to weed-killers, relates to the timing of application. Some herbicides, which would not be tolerated by the crops on which they are commonly used if they were applied after the crop had emerged, can be used selectively by incorporating them into the soil before the crop is planted, or by applying them to the soil surface after the crop is sown but before the germinating seedling breaks through the ground. These are known as pre-emergence herbicides. On the other hand, there are post-emergence herbicides, which, because of the selectivity of their action on different species of plant, can be used to control weeds at some stage after the shoots of the crop have emerged. Some herbicides can be used in both pre-emergence and post-emergence applications.

There is, in addition, another sense in which the terms 'pre-emergence' and 'post-emergence' are commonly used; this relates to the stage of growth of the weed rather than of the crop; as already mentioned under main uses of pesticides, herbicides which are taken up through the roots can be applied to the soil to control weeds before the weed seedlings have broken through the ground, whereas herbicides which are taken up through the leaves are effective only after the weeds have emerged.

An important distinction is that between systemic and non-systemic chemicals. A systemic pesticide is one which has the ability to penetrate the tissues of the plant (or animal in the case of a veterinary product) and, if fully systemic, to be carried (or translocated) in the transport system to parts remote from the site of application. Systemic activity can be important in a number of ways. For example, where a pest spends the whole, or the main part, of its life within the plant or animal, only a product which can penetrate the tissues can reach it.

Again, systemic action can assist the selectivity of a pesticide. For example, some species of insects, such as aphids and mites, feed by penetrating the plant surface with specialized mouth parts and sucking out the plant juices; a systemic product is particularly useful against such species and, if surface residues disappear rapidly, its use will result in minimum risk to other harmless organisms on the leaf surface. Systemic action also means that the whole surface of the plant or animal need not be exposed to the pesticide. This can be advantageous in reducing the volume of spray which would otherwise be needed to ensure thorough coverage and may simplify spraying techniques where the site of pest attack is difficult to reach directly.

*Activity, 'spectrum', and selectivity.* Toxicity is a subject that is too often thought of only in its negative sense. When it is applied to action on a pest, toxicity is the essential property of most pesticides; indeed, degree of toxicity to the pest is the basic criterion used to measure activity in the initial evaluation and selection (or 'screening') of chemicals as potential pesticides. The usual method of screening is to find out the amount of the chemical required to kill a certain proportion of a population of a particular species of organism, representative of a group of pests, in a given time and under carefully controlled conditions. This amount can then be compared with that of a known active ingredient (the 'standard'). The activity must later be confirmed against actual pest populations in the field — normally in field screening and field trials, which establish the dosage rates needed to achieve the required level of control, consistently and under a variety of practical conditions. The performance of a product in controlling pest populations in the field is known as its 'biological efficacy'.

The range of organisms controlled by a pesticide at the recommended dosage rate is referred to as its 'spectrum of activity'. A pesticide is classed as 'specific' where only a single species or well-defined class of organism is controlled; as 'narrow spectrum' where a relatively limited number of species are affected; or as 'broad spectrum' where a wide range of different species is controlled.

The degree to which different species are susceptible to the recommended dosage will also determine the pesticide's selectivity. In the case of weedkillers, selectivity usually refers to the ability to control certain weeds without damaging the crop; for other pesticides, it is the ability to control particular pests without adversely affecting harmless or beneficial organisms. Although specific and narrow-spectrum products are usually highly selective, they are not necessarily so; the distinction between the two terms is that 'spectrum' refers solely to the range of pests controlled without reference to other organisms.

While selectivity is obviously a desirable property, unfortunately most crops are attacked by a variety of different pests at the same time: this is called 'a pest complex'. It is frequently impracticable and uneconomic to control each pest with a different specific or highly selective pesticide, or to use mixtures of such pesticides. To control pest complexes, a broad-spectrum pesticide is highly advantageous, provided it is also selective to the crop; choice of pesticide and the method and timing of application can minimize the risk of any adverse effect on beneficial organisms.

Toxicity to plants is, of course, an important factor in the use of herbicides, and in the use of chemicals for getting rid of unwanted parts of the crop to facilitate harvesting (such as haulm killers, desiccants, and defoliants). On the other hand, toxicity and phytotoxicity (damage to plants) are also among the most undesirable properties of pesticides.

Pests belong to the animal kingdom or the plant kingdom just as do man himself, his crops, his pets and livestock, and the plants and creatures he finds beneficial or aesthetically pleasing. So it is not surprising that many of the most effective pesticides are also toxic to at least some of these desirable organisms and can be harmful if they are not handled with care.

Persistence is another property of pesticides which may be advantageous or disadvantageous. Applied to the length of time over which a pesticide will remain active in controlling a pest, persistence (more correctly referred to as 'residual activity') can be an important attribute, obviating the need for expensive and time-consuming re-treatment. But persistence is obviously undesirable where it is carried beyond the period of pest attack, especially where harmful concentrations are retained in food crops or in the environment.

*Toxicology.* The toxicity of a pesticide to organisms other than pests is investigated by exposing representatives of such groups of organisms to the compound (referred to as the 'toxicant') in experimental studies. The most important factor is that of mammalian toxicity, which covers both man and those domestic animals and pets which are mammals. The test species are normally rodents (particularly rats and mice), which are easily reared in the laboratory as a pure strain under standard conditions.

Toxicity is described as 'acute', 'sub-acute', or 'chronic', according to the level and duration of exposure. Acute toxicity, which is a measure of the hazard to man and animals following a single exposure to the pesticide, is usually expressed as the dose in milligrams of the chemical per kilogram of body weight of the test animal which will cause the death of a given percentage of the population. This is referred to as the lethal dose; thus, the $LD_{50}$ is the dose which will kill 50 per cent of a given population. Because different species, and even different sexes of the same species, can differ markedly in their susceptibility to a particular chemical, the lethal dose is qualified by reference to the particular species and sex of the test animal used.

The way in which the toxicant is administered, or in which the animal is exposed to it, also has an important and varied effect on its relative toxicity: for example, 'acute oral toxicity' is a measure of the lethal dose when the compound is given as a single dose by mouth; 'acute percutaneous toxicity' (sometimes loosely referred to as 'dermal toxicity') is used when the compound is applied to the skin; 'intraperitoneal toxicity', when it is injected into the body cavity; and 'inhalation toxicity' when it is breathed in as a vapour. The way in which a compound is administered is therefore a very important qualification necessary for a proper interpretation of tests. Again, the form in which the toxicant is administered — for example, as the pure material, the technical grade or as a solution in different solvents — can also profoundly affect the lethal dose.

It can be seen, then, that a great deal of detailed information is needed before valid conclusions can be reached about the mammalian toxicity of a compound. One chemical may be more toxic than another if taken orally, but their relative toxicities may be reversed if they are applied to the skin: toxicity studies on formulated material are likely to be more relevant to assessing the hazard to workers handling a pesticide in the field than are those on the 'technical material'.

Exposure to repeated sub-acute quantities gives evidence of cumulative toxicity; and if this is extended approximately over the life-span of the animal it is called 'chronic' toxicity. Measurement of these forms of toxicity is undertaken on test animals in feeding studies in which small dosages, usually expressed in parts per million (ppm) of the toxicant in the food, are fed over periods of up to two years. A range of dosages is used to establish the maximum dosage which will produce no detectable toxicological effect in the test animal (this is known as the 'no-effect' level.

Chronic toxicity tests are important to the safety of people who eat foods in which traces of the pesticide may remain. Residues of the pesticide in harvested crops are measured after field trials at various dosage rates, and at a range of intervals before harvest. Cumulative toxicity tests are important to people who work in the pesticide manufacturing plant or who apply the pesticide in the field.

Other tests are carried out to discover whether there will be any undesirable effects on fish, wild life, or other beneficial organisms; and there are carcinogenicity and teratogenicity studies to establish that the material does not cause cancer or abnormal growth of the foetus.

The results of all these toxicological studies will determine whether products containing the active ingredient can safely be recommended for use and, if so, what precautions are necessary and should be indicated on the label to ensure that the materials are handled and applied correctly. In most countries such decisions are controlled by government departments, to which all data must be submitted as part of 'regulatory' procedures whereby products must be registered before they are approved for sale (see Chapter 19).

**Reference**

1. MELLANBY, K. *Pesticides and pollution.* Collins (1967).

Table A.1. *Classification of pest-control agents*

| MAIN GROUP | SUB-GROUP | EXAMPLES |
|---|---|---|
| **INSECTICIDES** | | |
| INORGANIC | | aluminium phosphide, calcium arsenate |
| BOTANICAL (plant extracts) | | nicotine, pyrethrin, rotenone |
| ORGANIC | Hydrocarbon oils | citrus spray oils, dormant sprays ('winter washes'), mosquito larvicides |
| | Chlorinated hydrocarbons (Organochlorines) | aldrin, BHC, DDT, heptachlor, toxaphene |
| | Organophosphates —non-systemic | azinphos methyl, dichlorvos, ethyl and methyl parathion, fenitrothion, malathion |
| | — systemic | demeton methyl, dimethoate, monocrotophos, phosphamidon |
| | Carbamates —non-systemic | carbaryl, methomyl, propoxur |
| | — systemic | aldicarb, carbofuran |
| | Synthetic pyrethroids | allethrin, bioresmethrin, permethrin |
| MICROBIAL | Bacterial | *Bacillus thuringiensis* |
| | Viral | polyhedral viruses |

Table A.1. *Classification of pest-control agents (contd)*

| MAIN GROUP | SUB-GROUP | EXAMPLES |
|---|---|---|
| **OTHER INSECT-CONTROL AGENTS** | | |
| CHEMOSTERILANTS | | apholate, metepa, tepa |
| PHEROMONES (sex attractants and synthetic lures) | | Gyplure, Medlure, Siglure, Trimedlure |
| REPELLENTS | | deet, dimethyl phthalate, ethyl hexenediol |
| INSECT HORMONES AND HORMONE MIMICS (insect growth regulators) | Juvenoids (juvenile hormone mimics) | farnesol, methoprene |
| | Moulting inhibitors | diflubenzuron, ecdysone |
| **SPECIFIC ACARICIDES** | | |
| Non-fungicidal | Chlorinated hydrocarbons (organochlorines) | chlorobenzilate, dicofol, tetradifon |
| | Organo-tins | cyhexatin |
| Fungicidal | Dinitro compounds | binapacryl, dinocap |
| | Other | chinomethionat |
| **PROTECTANT FUNGICIDES** | | |
| INORGANIC | | bordeaux mixture, copper oxychloride, sulphur |
| ORGANIC | Dithiocarbamates | mancozeb, metiram, propineb, thiram, zineb |
| | Phthalimides | captafol, captan, folpet |
| | Dinitro compounds | binapacryl |

Table A.1. *Classification of pest-control agents (contd)*

| MAIN GROUP | SUB-GROUP | EXAMPLES |
|---|---|---|
| | Organomercurials | phenyl mercury acetate and chloride |
| | Organo-tin compounds | fentin acetate and hydroxide |
| | Others | chinomethionate chlorothalonil, dichlofluonid dichlone, dicloran, dodine, dyrene, glyodin |
| **ERADICANT FUNGICIDES** (chemotherapeutants) | | |
| | Antibiotics | blasticidin, cyclohexamide, kasugamycin, streptomycin |
| | BCM generators | benomyl, thiabendazole, thiophanate-methyl |
| | Morpholines | dodemorph, tridemorph |
| | Formylamino compounds | chloraniformethan, triforine |
| | Others | ethirimol, oxycarboxin |
| **SOIL FUMIGANTS AND NEMATICIDES** | | |
| SOIL STERILANTS | Halogenated hydrocarbons | chloropicrin, methyl bromide |
| | Methyl isothiocyanate generators | dazomet, metham |
| | Others | carbon disulphide, formaldehyde |

Table A.1. *Classification of pest-control agents (contd)*

| MAIN GROUP | SUB-GROUP | EXAMPLES |
|---|---|---|
| FUMIGANT NEMATICIDES | | |
| | Halogenated hydrocarbons | DD, dichloropropene, ethylene dibromide |
| NON-FUMIGANT NEMATICIDES | Organophosphates | dichlofenthion, fensulfothion, phenamiphos |
| | Carbamates | aldicarb, carbofuran |
| **HERBICIDES** INORGANIC | | sodium arsenite, sodium chlorate |
| ORGANIC | Phenolics | bromofenoxim, dinoseb acetate, DNOC, nitrofen, PCP |
| | Phenoxyacids ('hormone weedkillers') | CMPP, MCPA, 2,4-D, 2,4,5-T |
| | Carbamates | asulam, barban, benthiocarb, carbetamide, chlorpropham, phenmedipham, propham, triallate |
| | Substituted ureas | diuron, fluometuron, linuron, metobromuron, monolinuron |
| | Halogenated aliphatics | dalapon, TCA |
| | Triazines | ametryne, atrazine, methoprotryne, simazine, terbutryne |
| | Diazines | bromacil, lenacil, pyrazon |

Table A.1. *Classification of pest-control agents (contd)*

| MAIN GROUP | SUB-GROUP | EXAMPLES |
| --- | --- | --- |
| | Quaternary ammonium compounds | |
| | — bipyridyls | diquat, paraquat |
| | — pyrazolium | difenzoquat |
| | Benzoic acids | chlorfenprop methyl, dicamba, 2,3,6 - TBA |
| | Arsenicals | cacodylic acid, DSMA, MSMA |
| | Dinitroanilines | nitralin, profluralin, trifluralin |
| | Benzonitriles | bromoxynil, chlorthiamid, dichlobenil, ioxynil |
| | Amides and anilides | benzoylprop-ethyl, diphenamid, propachlor, propanil |
| | Others | aminotriazole, flurecol, glyphosate, picloram |
| **DESICCANTS, DEFOLIANTS, HAULM KILLERS\*** | | |
| | Quaternary ammonium compounds (bipyridyls) | diquat, paraquat |
| | Phenolics | cacodylic acid, dinoseb, DNOC, PCP, sodium chlorate |

\*Although compounds in this category are also herbicides, they are here being used on the crop itself, and in such applications are sometimes included in the general term 'plant growth regulators'.

Table A.1. *Classification of pest-control agents (contd)*

| MAIN GROUP | SUB-GROUP | EXAMPLES |
|---|---|---|

**PLANT GROWTH REGULATORS (PGR's)**

| MAIN GROUP | SUB-GROUP | EXAMPLES |
|---|---|---|
| GROWTH PROMOTANTS<br>(auxins and auxin type) | | gibberellic acid |
| GROWTH INHIBITORS<br>(stem shorteners) | Quaternary<br>ammonium<br>compounds | chlormequat |
| SPROUT INHIBITORS<br>AND DESUCKERING AGENTS | | |
| Herbicidal | Carbamates | chlorpropham,<br>propham |
| Specific | | maleic hydrazide,<br>'Offshoot T' (mixture<br>of fatty alcohols) |
| FRUIT SETTING,<br>RIPENING, FLOWERING<br>AGENTS AND LATEX<br>STIMULANTS | Ethylene generators | ethephon |
| | Others | dimas, glyphosine,<br>naphthaleneacetic<br>acid |
| FRUIT DROP<br>INDUCTION<br>(Abscission agents) | | cycloheximide |

**RODENTICIDES**

| MAIN GROUP | SUB-GROUP | EXAMPLES |
|---|---|---|
| FUMIGANTS<br>(space fumigants also used<br>for rodent control) | | aluminium phosphide<br>calcium cyanide,<br>chloropicrin, methyl<br>bromide |
| ANTI-COAGULANTS | Hydroxy coumarins | coumatetralyl,<br>difenacoum, warfarin |
| | Indandiones | chlorophacinone,<br>phenyl-methyl<br>pyrozolone, pindone |

Table A.1. *Classification of pest-control agents (contd)*

| MAIN GROUP | SUB-GROUP | EXAMPLES |
|---|---|---|
| OTHERS | Arsenicals | 'arsenious oxide', sodium arsenite |
| | Thio-ureas | antu, promurit |
| | Botanical | red squill, strychnine |
| | Others | norbormide sodium fluoroacetate, vitamin D (calciferol), zinc phosphide |
| **MOLLUSCICIDES** | | |
| AQUATIC | Botanical | Endod |
| | Chemical | copper sulphate, niclosamide, sodium pentachlorophenate, trifenmorph |
| TERRESTRIAL | Carbamates | aminocarb, methiocarb, Zectran |
| | Other | metaldehyde |

# Index

SOCIAL SCIENCE LIBRARY
...sity Libr...